Agrarian Reform in Theory and Practice

A study of the Lake Titicaca region of Bolivia

JANE BENTON

Routledge
Taylor & Francis Group

LONDON AND NEW YORK

First published 1999 by Ashgate Publishing

Reissued 2018 by Routledge
2 Park Square, Milton Park, Abingdon, Oxon, OX14 4RN
711 Third Avenue, New York, NY 10017, USA

Routledge is an imprint of the Taylor & Francis Group, an informa business

Publisher's Note
The publisher has gone to great lengths to ensure the quality of this reprint but points out that some imperfections in the original copies may be apparent.

Disclaimer
The publisher has made every effort to trace copyright holders and welcomes correspondence from those they have been unable to contact.

A Library of Congress record exists under LC control number: 99076346

ISBN 13: 978-1-138-61009-5 (hbk)
ISBN 13: 978-1-138-61010-1 (pbk)
ISBN 13: 978-0-429-45943-6 (ebk)

Contents

Maps vi
Photographs vii
Preface viii
Acknowledgements x

Part I: The Theory
1 Agrarian reform in Latin America 1
2 Agricultural systems in pre-Columbian Bolivia 15
3 The domination of the *hacienda* 27
4 Bolivia's Agrarian Reform Law of 1953 45
5 *Ley INRA* (1996) 69

Part II: The Practice
6 The Lake Titicaca region 97
7 The state of agriculture in lakeside communities on the eve of agrarian reform 113
8 The impact of the 1953 agrarian reform legislation 131
9 *Campesino* opinions on agrarian reform 149
10 Some conclusions to be drawn from the case study 167
11 The crisis in *campesino* farming in the Lake Titicaca region at the end of the twentieth century 181

Abbreviations used in the text 193
Glossary 195
Bibliography 200

Maps

Bolivia: Political Regions   16

Bolivia: Major Physical Regions   24

Lake Titicaca   98

Part of the Lake Titicaca Region of Bolivia   104

Copy of plan of Chua Visalaya (1957)   136

Photographs

1   Aerial photograph of Aymara farming communities bordering the north-eastern shore of Huiñaymarca: from left to right, Compi, Llamacachi, Chua Visalaya and Chua Cayacota (Instituto Geográfico Militar, La Paz, 1956)

2   Some of Llamacachi's lakeside fields, planted with potatoes, onions and beans (1971)

3   The effects of *El Niño* flooding in Llamacachi (1987)

4   Pre-Columbian walled terraces near Copacabana (1981)

5   Sheep grazing on former *hacienda* pastures in Chua Visalaya, with disused hillside terraces (1987)

6   Chua Visalaya's derelict *hacienda* house and abandoned agricultural machinery (1971)

7   Members of Chua's cooperative ploughing (1971)

8   *Campesinos* ploughing and planting potatoes near Huatajata (1991)

9   Members of Compi's farming syndicate marching to a lakeside *fiesta*, led by a bedecked tractor, the 'main guest' (1991)

10 Potatoes and *oca* being dehydrated in Chua (1996)

11 The weekly (Thursday) market in Jank'o Amaya (1996)

12 Indigenous groups setting out from Santa Cruz on the *Ley INRA* march to La Paz (August 1996)

13 The 'siege of La Paz': *campesinos* and coca growers protesting about the proposed *Ley INRA* (September 1996)

14 Llamacachi's first hotel, built by a Paceño on a plot of land purchased from a rural-urban migrant (1998)

15 Tourists visiting the reed boat museum near Huatajata (1998)

Preface

Several colleagues have questioned the point of writing a book on agrarian reform at the turn of the century, commenting that both land and agrarian reform have been 'dead issues' for at least two decades. Whilst it is true that a significant number of Third World countries embarked on ambitious reform programmes in the 1960s and 1970s, to numerous governments the goal of sustainable agriculture has never appeared more elusive. Likewise, 'the land question' featured prominently on the agenda of the 1992 Rio Earth Summit. As I write these lines, land reform is being vigorously contested in countries as far apart as Brazil, Zimbabwe and Scotland.

This book is the outcome of my fascination with Bolivia, especially with Aymara farming communities in the Lake Titicaca region, for more than three decades. I first visited Bolivia with a friend in November 1966 (our arrival in the country coinciding with that of Che Guevara), having travelled 'roughly' for several months after completing a Voluntary Service Overseas project and celebrating independence in Guyana. Whilst carrying out research in 1971 for my doctoral thesis, I had the privilege of living for six months with my Aymara field assistant's family in Llamacachi, a small farming community beside the lake. Since then I have made five return visits (in 1981, 1987, 1990, 1991 and 1996) to the same region, in the course of researching on agricultural change, community development, the role of non-governmental organizations and the activities of women's grassroots groups. In 1998 I visited the country for the eighth time as a participant in a British Council-funded Glasgow University Geography Department academic link with higher education institutions in La Paz.

Most of the books, papers and articles on Bolivia's agrarian reform have been written by land economists (notably by academics based at Wisconsin University Land Tenure Center), agrarian analysts and political economists. As a social geographer my approach and interests are somewhat different: in compiling the book I have been concerned primarily with the interactions between human communities and the physical environment, with various aspects of agricultural and rural change and with the perceptions and opinions - rarely sought by government personnel or even researchers - of individuals, families and farming communities directly affected by the processes of change.

Whilst an area dominated by small peasant farming communities, adjacent to the world's highest commercially navigable lake, may seem remote from the experiences of a British academic, I have never found this to be the case: indeed, I have come to regard it almost as a second home. Most of my life has been spent in what was until the 1920s a village estate in the ownership of a wealthy banking family. On one fateful day in February 1921 'land reform' occurred in dramatic style! All estate properties, with the exception of one farm, were sold by auction in lots, mainly to the former tenant farmers and farm workers who could afford to buy them. Until the late 1950s the village remained virtually intact as a small farming community in central England. Remarkably the comments of secondary school children at that time, about not intending to become farm labourers like their fathers, but wanting to work instead in a factory or move to a big city, were almost identical to those expressed by teenagers living in communities by Lake Titicaca in the 1970s and 1980s. Over the last 30 years my home village has gradually been transformed into a commuter settlement, retaining less than a handful of the original farms and smallholdings. Recent visits to Bolivia have convinced me that comparable changes are beginning to take place in those lakeside communities which enjoy easy access to La Paz.

The book falls naturally into two parts. After a general introduction to agrarian reform in Latin American republics, agricultural changes in Bolivia (including what some agrarian analysts regard as colonial agrarian reform prototypes) are traced from pre-historic times up to the radical 1952 National Revolution. The fourth chapter reviews the deliberations preceding, and the contents of, the 1953 agrarian reform legislation. The final chapter of Part I is devoted to the recently enacted, highly controversial *Ley INRA*, variously referred to as Bolivia's 'modified', 'new' or 'second' agrarian reform law, although it will become apparent that it is more aptly termed a land reform act. In Part II the difficulties of implementing agrarian reform policies and their impact on the countryside are examined in the context of the Lake Titicaca region. Lakeside dwellers' viewpoints on landownership, agrarian reform and rural change in general are considered in the ninth chapter. Chapter 10 highlights some of the shortcomings of Bolivia's agrarian reform policies, in addition to the human conflicts associated with land redistribution. In the final chapter evidence is presented in support of my conviction that lakeside farming has reached a critical stage and peasant farming communities face a bleak, uncertain future.

Acknowledgements

My thanks are due to Lindsay Stout, without whom I might never have visited Bolivia, and to Sofia Velásquez Laura, without whose assistance and friendship I would have found it difficult to complete my thesis and to gain insights into Aymara agricultural practices and relationships with the natural and supernatural worlds. I would also like to thank Professor C.T. Smith for reading and commenting on the first chapters and David Fox for suggesting that I write the book in the first place. Additionally I must thank Andrew Lawrence of the Keele University Cartographic Unit for preparing the five maps and Sue Allingham for typing the text. I am grateful to Professor Arthur Morris, organizer of the British Council academic link referred to in the Preface, for his support and to my family, friends and colleagues for their encouragement whilst writing the book.

April 1999   Jane Benton

Part I:
The Theory

1 Agrarian reform in Latin America

At the end of the twentieth century Latin America's 'land question' poses as great a problem and arouses as much antagonism as it did more than 80 years ago, when Mexico embarked on the region's first radical agrarian reform programme. Despite numerous and widespread attempts to implement land and agrarian reform policies, land remains a vital social, economic and political issue in virtually all Latin American republics. Indeed, it has even been claimed that 'land is the most important political issue in Latin America today, and unequal landownership the greatest cause of poverty and inequality in the societies of the continent' (Gelber, 1992, p.129).

The struggle for access and legal rights to land resources continues to provoke violent unrest: in some countries the rural poor are still literally 'dying for land'. In recent years land disputes have precipitated civil warfare in the Central American republics of El Salvador, Guatemala and Nicaragua. Brazilian 'landless peasants and landowners ... say the fight over land has created a climate of war' in the country (*Latinamerica Press*, 1997). The outside world was shocked in April 1996 when a peaceful road blockade demonstration about landownership rights in the Brazilian state of Pará turned into a massacre by the police, leaving 19 landless rural labourers dead and more than 40 wounded. Several months later more than 30,000 Bolivian *campesinos* (countrymen, usually denoting peasant farmers), *cocaleros* (coca producers) and *pueblos indigenas* (indigenous peoples) participated in a month-long march for 'Territory, Land, Political Rights and Development', in protest against some of the clauses of the government's proposed agrarian reform law. What was described in national newspapers as 'the march of the century' gave way to a period of two weeks during which La Paz was 'besieged' and traffic frequently brought to a stand-still by the participants.

Such public demonstrations and land invasions are only to be expected in countries with tremendous inequalities in terms of landownership and wealth; where governments have traditionally placed a very low priority on agricultural development, reflected in minimal financial investment; where food production has failed miserably to keep pace with population growth and where excessive land fragmentation and landlessness have contributed in large measure to severe land degradation, malnutrition and mass rural-urban migration. 'According to the Inter-American Development Bank, the wealthiest 10 per cent of Latin Americans receive 40 per cent of the region's income, while the poorest 30 per cent share only 7.5 per cent of the income ... About 150 million Latin Americans subsist on less than US $ 2 per day' (*LP*, 1999).

Definitions of land and agrarian reform

Inevitably confusion and misunderstandings over definitions arise when the terms land reform and agrarian reform are used loosely and interchangeably. The problem is particularly complicated in Latin America, where the single term, *reforma agraria*, applies to both land and agrarian reform.

Whilst land reform is generally recognized by politicians, economists and agrarian analysts as a major form of state intervention in rural affairs and, by some, as an essential component of Third World economic development, its potential per se. for raising agricultural productivity and promoting sustainable development is limited. All-encompassing agrarian reform measures, rather than changes in access to land alone, are normally indicated when such weighty claims are advanced. Commission and conference resolutions tend to be more precise in their usage of terminology. Thus the recommendations of the 1979 World Conference on Agrarian Reform and Rural Development emphasized that 'agrarian reform is a critical component' of rural development. The following year Brandt Commissioners reported that: 'Agrarian reform is a critical means to benefit the poor - though naturally the measures needed differ from country to country' (Brandt, 1980, p.95).

Writing in 1971 (p.177), Feder acknowledged that 'there is not complete unanimity with respect to the meaning, nature and scope of land reform. In fact, the more the issue is debated, the more confused people are likely to become'. Noting that 'land reform has fascinated politicians and political philosophers from time immemorial', Edmundo Flores, the Mexican agrarian expert who acted as adviser to the Bolivian government in the early 1950s, stressed the political aspects: he described land reform as 'a revolutionary measure which passes power, property and status from one group of the community to another' (1970, p.151). A more comprehensive definition was given in a 1974 World Bank Paper: 'Land reform involves intervention in the prevailing pattern of land ownership, control, and usage in order to change the structure of holdings, improve land productivity, and broaden the distribution of benefits'.

Today there is a general consensus on the broad definition of land reform. According to Thiesenhusen (1995, p.6), 'land reform is any fundamental alteration of the existing tenure, usually understood to mean redistribution of tenure rights from one group (usually elite landowners) to others (usually peasants without land or with insecure access to it)'. Abbott and Makeham (1990, p.168) view land reform as 'a fundamental change in institutionalized relationships among people with respect to land'. The definition used by Dickenson et al. (1996, p.145) refers to different types of land reform:

> Land reform is understood to mean the reorganization of landholding and tenurial structures which may involve: (a) the expropriation of large estates, with or without compensation, and their reorganization into peasant farms, cooperatives or collective farms so as to benefit peasants and landless labour; or (b) the consolidation of very small or fragmented holdings into holdings of an adequate size.

Definitions tend to gloss over the tenancy component of land reform. As Barke and O'Hare (1991, p.104) state, 'tenancy reform ...usually involves improvements in tenancy contracts such as providing more security of tenure, introducing more equitable crop sharing and rent arrangements'. The Brandt Report (1980, p.95) recognized that: 'In some areas the key issue is reform of tenancy to give greater security of tenure. In others it is to divide large parcels of land among those who can farm it more intensively. Yet others require consolidation measures to overcome the effects of excessive fragmentation of holdings which has already occurred'.

Whilst some Latin American republics, including Bolivia, have incorporated planned colonization projects in their land or agrarian reform programmes, a number of land economists have from the 1960s been strongly opposed to the inclusion and implementation of such measures; of recent years their arguments have been reinforced by those of social anthropologists and environmentalists. Flores (1970, p. 149) was adamant that:

> Land reform ... should not be confused with attempts to reclaim unproductive land or settle uninhabited areas ... We should not forget that in the course of several centuries those lands failed to tempt either the Indian farmers who preceded Columbus, the Spanish conquerors or the Catholic Church - all of whom coveted land and knew what to do with it.

Thiesenhusen (1995, p.13) maintains that colonization in what are usually 'ecologically fragile areas ... except in relation to indigenous peoples (who live there already), is usually environmentally dangerous and should be halted'.

Despite such warnings, there has been a growing tendency during recent decades for governments to opt for colonization projects, thereby placating politically organized peasants and socially aware middle class urban dwellers, whilst simultaneously diverting attention from uncultivated or under-cultivated, privately owned, extensive landholdings. In the case of Venezuela, the land selected by the government for 'agrarian reform' colonization schemes was already occupied and had to be purchased for distribution to peasant farmers. 'The price paid to the landowners was extremely favourable and this 'reform' has been seen as little more than a bonus for the landowning class and a means of payment to the peasant clients of the two main political parties' (Whittemore, 1981, p.22).

Agrarian reform generally embodies land reform as its most important element. 'Agrarian reform includes both redistributing land and assisting new landowners by assuring them inputs and markets, extending credit and imparting certain technology that will help them to become agricultural producers' (Thiesenhusen, 1995, p.12). A carefully planned agrarian reform law 'includes all the necessary institutional reforms which influence or relate to the use of land' (Whittemore, 1981, p.12). Of prime importance is the provision of rural credit and extension services. Peasants or peasant cooperatives in possession of newly acquired titled lands can accomplish little, if anything, without access to adequate financial resources, enabling them to purchase seeds, stock, tools, fertilizers, pesticides etc. and, in some cases, install water supplies. 'Agricultural extension assists farm people through educational procedures, in improving farming methods and techniques, increasing production efficiency and income, bettering their levels of living and lifting the social and educational standards of rural life' (Alexandratos, 1995, p.344). The creation or extension of marketing facilities and improvements to transport systems are also essential to alleviating rural poverty and raising farm incomes above bare subsistence levels.

The World Bank agrees that agrarian reform 'is a more comprehensive concept than land reform, since it involves a wide modification of a wide range of conditions that affect the agricultural sector'. However the 1974 Paper (p.18) added the highly significant point, sometimes forgotten, that land reform is not an inevitable component of agrarian reform:

Agrarian reform may or may not include land reform; in some instances there may be no need for land reform since land is already evenly distributed. In other cases, it may not be politically feasible to have land reform ... though it might be both politically and economically feasible to raise output through the measures involved in agrarian reform.

4

The same document summarized under four headings the World Bank's general conclusions about land and agrarian reform, based on its experiences of funding programmes and projects in the 1960s and early 1970s. Firstly, there is a recognition of the 'overriding importance of the political factor' in restructuring landholding systems, since 'a meaningful land reform program will inevitably destroy or limit the power base of many persons'. Secondly, the establishment of national institutions and rural organizations are crucial to the successful implementation of land reform: otherwise, collusion between landowners and unsupervised officials, 'combined with an absence of organized pressure from the beneficiaries' can nullify 'positive reform efforts'. Thirdly, reforms are rarely carried out 'without considerable upheaval and loss of production' - a point discussed later. Finally, perspectives on the effectiveness of reform change over time: a number of positive effects may not be apparent for some years, whilst the withdrawal of support systems (e.g. rural credit facilities and agricultural extension services) may reverse initial gains.

Latin America's 'land question'

Whilst 'the debate over theories of development has been an active one in Latin America, and...has greatly affected the way reform issues are formulated and the means by which reform objectives are pursued' (Dorner, 1992, p.14), land and agrarian reform programmes have generally been introduced for one or both of the following basic reasons: (i) a desire for social justice and (ii) the need to increase economic efficiency with the aim of raising agricultural productivity.

Although, since the turn of the century, many Latin American governments have preferred to ignore the social injustices of their countries' landholding structures, few have denied their existence. According to Flores (1970, p.157), when land reform began in Mexico in 1917, 'less than 3 per cent of the landowners together owned over 90 per cent of the productive land'. Slight discrepancies in statistical data can in no way detract from the gross inequalities of landownership at the point when the revolutionary government embarked on what has been referred to as 'the most violent' and 'most bloody' of the early agrarian reforms. Throughout Latin America land and power were - and in a number of republics, still are - concentrated in the hands of a small, privileged minority of landlords. Vast *haciendas* (landed estates, traditionally cultivated by a system of serfdom), usurped from peasant communities during the centuries following the Spanish Conquest, ruthlessly exploited landless and land-poor tenants, granting them usufruct rights to minuscule plots of land. In return, families were obliged to carry out tedious, often back-breaking, 'feudal' field tasks and domestic duties. Such domination and exploitation 'was legitimised by the

Roman Catholic Church, which, during the colonial period, was itself a vast landholder' (Whittemore, 1981, p.6).

In countries such as Brazil, Paraguay, Uruguay and Panama, flagrant inequalities persist today. Salles (*LP*, 1994) claims that in Brazil 'the land concentration is so severe that 501 landowners own 57 million hectares'. Vidal (1997) describes the country's landholding structure in even more dramatic terms:

> Today, Brazil has a population of 165 million, yet fewer than 50,000 own most of the land in a country 66 times larger than Britain. At the other end of the scale, four million peasant farmers share less than 3 per cent of the land. Moreover, some 42 per cent of all privately-owned land in Brazil lies idle, not even grazed by cattle.

As members of *Sem Terra* (Without Land - a movement of croppers, casual pickers, farm workers and the victims of land clearances) testify, 'a great mass of landless families work as farm labourers at brutally low wages' (Salles). The implications of the age-old dichotomy of landownership in Latin America are aptly summed up by Gelber (1992, p.129):

> The special tensions of landownership in Latin America come from the juxtaposition of two extremes: large numbers of rural people live in deep poverty, with little or no land, while ownership of huge holdings is concentrated in the hands of a tiny minority, creating a degree of inequality not matched in any other region of the world.

Since Bolivia's agrarian reform programme was launched in 1953, as one of the major outcomes of a thorough-going national, social revolution, there has been throughout Latin America 'a rising groundswell of protest from peasant populations because of the increasing fragmentation of tiny holdings, encroachment by landowners on common lands or their exploitation of tenants and labourers' (Dickenson et al., 1996, p.146). In recent years politically organized groups of *campesinos* have received welcome support from diverse quarters viz. rural-urban migrants, informed, socially concerned middle class city dwellers, unions and both national and overseas non-governmental organizations (NGOs). Brazil's *Sem Terra*, now a highly organized mass movement of rural activists, intent on securing social justice through agrarian reform (and using slogans such as 'Agrarian Reform! The battle is ours!'), has earned world-wide acclaim for its well planned occupations of uncultivated estate lands: Noam Chomsky even suggests that it 'may be the most important grassroots social movement in a world where the left is deeply confused about direction and path' (Vidal, 1997).

Over recent years the Roman Catholic Church has become increasingly involved in Latin America's land debate. In April 1997 Guatemala's Chamber of

Agriculture accused a bishop and four priests of inciting landless peasants to invade estate lands in the northern department of El Petén. Nine months later such land occupations were even given the Vatican's qualified approval. In a papal document entitled, *For a better land distribution: The Challenge of Agrarian Reform*, it was acknowledged that 'land takeovers are not in accord with civil law, but they are a way of making land reform come about more rapidly' (*LP*, 1998).

Additionally, in a number of Latin American states tribal indigenous peoples, whose territorial rights have until recently been largely disregarded in agrarian reform legislation, are joining forces and engaging in a desperate struggle for survival against powerful transnational logging, mining and oil interests. As Chauvin (*LP*, 1997) observes: 'Although the conflict between natural resource exploitation and indigenous peoples is not new, the new globalized economy and free-market policies followed in Latin America, have made it easier for the transnationals to make headway in the region'. Natural resources are being sold wholesale, regardless of the devastation to indigenous peoples, their land and their environment. The significance of land titles has been forcibly demonstrated in recent legal decisions brought to the public's attention by the media and the NGOs. For example, in September 1996 Nicaragua's forest-dwelling Sumu people went to the Appeals Court to ask the judges 'to block the government's concession of 63,000 hectares of forest' (their traditional homeland) 'to a lumber subsidiary of a Korean company'. The petition was rejected on the grounds that: 'No one lives there ... Until someone can show me a land title, these lands belong to the state' (Chauvin).

The granting of territorial land rights, the allocation of plots of cultivable land with titles, security of tenure and the abolition of *hacienda* obligations and services are generally regarded in Latin America as the major social gains from land reform. According to the Charter of Punta del Este (1961), the legal framework of the Alliance for Progress, 'the land will become for the man who works it the basis of his economic stability, the foundation of his increasing welfare and the guarantee of his freedom and dignity' (Feder, 1971, p.185). In Bolivia - as throughout Latin America - 'the land belongs to him who works it' has over the years become an extremely powerful and emotive slogan, uniting *campesinos* in anti-authority protests. Significantly, the August 1990 land rights march of Bolivia's eastern lowland indigenous peoples was not solely for legal rights to land: some of the marchers carried banners for more than 650 km bearing the words *Territorio y Dignidad* (Territory and Dignity). It will be argued later that *dignidad de la persona* (personal dignity) represents as valued an acquisition to the rural poor as title deeds to land.

The early agrarian reform laws of Mexico (1917), Bolivia (1953) and Cuba (1959) were regarded within those countries as integral components of social revolutionary government programmes. In Cuba agrarian reform was 'part of a process that brought almost the entire economy under state control' (Gelber, 1992, p.139). Social justice for the rural poor and increased economic efficiency were viewed in each country as inextricably linked driving forces behind legislation designed to transform an antiquated, inflexible land tenure system and 'modernize' agriculture. The close ties between social and economic motives were stressed by the economist, Raúl Prebisch (1962), in his role as executive secretary of the Economic Commission for Latin America (ECLA):

> The system of land tenure that still prevails in most Latin American countries is one of the most serious obstacles to economic development...The land tenure system is characterized by extreme inequality in the distribution of land and of the income accruing from it ... Redistribution of the land pursues the following two basic objectives: to relieve social tensions by improving the distribution of property and income; and to increase productivity by creating conditions favourable to the introduction of modern techniques (quoted in Stavenhagen, 1970, p.107).

Approaches to agrarian reform changed markedly during the 1960s. Of the 19 Latin American governments introducing reform measures in response to a call by the Alliance for Progress for 'social reform', few appear to have been motivated by a sincere desire to eradicate social inequalities and improve the lot of the rural poor and landless. Undoubtedly, the majority were primarily concerned about stemming the tide of potentially destabilizing *campesino* protest and establishing favourable conditions for overseas investment. There is no lack of evidence to support Gelber's allegation (1992, p.136) that: 'While Latin American governments were quite happy to take US money to set up land-reform institutes, they deliberately avoided confrontation with large landowners, who vehemently opposed any curtailment of their privileges and property rights'.

As the example of Chile dramatically demonstrates, the powerful combined opposition of large landowners, businessmen and the military to radical agrarian reform measures can lead ultimately to a total rejection and reversal of former policies i.e. to counter-agrarian reform. Under the Frei government, 3.5 million hectares of land were expropriated between 1967 and 1970 and subsequently transformed into *asentamientos* (cooperative farms). The Allende government extended land redistribution policies and succeeded in virtually abolishing *latifundio* properties (extensive, unproductive estates) by reducing the maximum size of holding to 40 hectares and expropriating a further 6.5 million hectares. The military regime replacing Allende's Popular Unity government in 1973 deliberately set out to reverse the process of agrarian reform: counter-reform measures returned almost 30 per cent of the land previously expropriated to the former *hacienda* owners. According to Kay and Gwynne (1997, p.4), 'a large proportion of land reform beneficiaries were left without land and over the years

8

over half of the *parceleros'* (people with small plots of land) 'sold their land as they could no longer afford running their family farms'. As Dorner (1992, p.39) remarks, 'the belief that land reform had negative impacts on agricultural production was a major justification given by the military government for turning back the reform, yet there is no evidence to support this belief'.

It is generally postulated that the acquisition of titles to land and security of tenure enable individual peasant farmers to work much more efficiently within an improved social and economic framework. They are given the incentive to make additional efforts in order to achieve greater productivity from which they, rather than their former employers, will reap the financial rewards. Without security of tenure 'land users apply only those inputs bringing immediate benefit' (Abbott and Makeham, 1990). As owner-occupiers of smallholdings they become more concerned about crop yields, improving animal husbandry, maintaining soil fertility, conserving water supplies, obtaining expert advice on farming techniques, introducing new forms of technology etc.

Such claims have sometimes been extended to peasant farmers working together in cooperatives or on collective farms. A number of Latin American agrarian reforms have advocated the formation of producer cooperatives, according to Dorner (1992, p.7), 'because of capital infrastructures and underdeveloped entrepreneurial skills among the potential beneficiaries of land redistribution'. Andean governments have espoused cooperatives convinced that, because Inca and Aymara Indians farmed cooperatively in pre-Conquest times, their descendants would be eager to avail themselves of the new opportunities offered. Cooperation was a central theme of Bolivia's 1953 legislation and is discussed later. Regrettably, in a number of cases cooperatives have met with little success. 'Many cooperatives in Peru sank under the weight of peasant insistence on individual holdings' (Dickenson et al., 1996, p.149): according to Alexandratos (1995) Peruvian cooperatives 'suffered from serious diseconomies of scale and work incentive problems and many were broken up in the early 1980s'. Collective farming in Mexico has suffered a similar fate. Very few of the *ejidos* (collective farm units) created as a result of agrarian reform legislation are today being worked on a collective basis; the majority are 'parcelled', with plots cultivated exclusively by individual families. Thiesenhusen (1995, p.41) attributes the breakdown of collective farming and *ejido* organization partly to 'factionalism and class consciousness' but also to the fact that the most skilled individuals 'tended to find individual farming ... more remunerative'.

Institutional economists have since the early 1970s argued the case for land reform on the grounds of efficiency, alleging that there is an inverse relationship between farm size and productivity. Output per unit of land is said to be higher on smallholdings - assuming they are 'large enough to provide the operators with satisfactory income under the prevailing conditions' (Abbott and Makeham, 1990, p.170) - than on large landholdings, because smallholders utilize a higher

9

proportion of cultivable land and generally, unlike estate owners, employ unpaid labour. Thus it is claimed, allocation of expropriated, under-used estate land to land-poor and landless peasant farmers should stimulate higher productivity. At the same time it has been recognized that overall production levels could fall initially, as adjustment to structural change occurs. Additionally, cases of increased peasant consumption of food, after years of 'imposed' malnutrition, have been recorded in association with a number of agrarian reforms.

According to this theory, distribution of land to *campesinos* would set in motion a number of structural economic changes: 'the creation of relatively small family farms (or cooperative forms of tenure) would provide employment for excess labor, relieve the pressures of urban migration, and release labor, in a more controlled and beneficial way' (Dorner, 1992, p.18). Rising rural incomes would generate increased demands for industrial goods and the ultimate outcome would be 'an integrated process of agricultural-industrial development'.

Notwithstanding, the inverse relationship theory appears extremely idealistic and simplistic in today's world of harsh realities. It makes a number of unrealistic assumptions and calls for careful application at local, regional and national levels. At all events, how is one able to justify arguing the case for economic efficiency on the grounds that costs are minimized because peasant labour is either unwaged or very poorly paid? Moreover, a poverty-stricken rural population can hardly be expected to create increasing demands for industrial goods. There is an underlying assumption that government support, in the form of rural credit facilities and extension services, is readily at hand and will continue to be forthcoming, to enable small farmers to become increasingly efficient. In some Latin American countries *campesinos* **did** receive various forms of government assistance in the early days of agrarian reform but in most cases such support was withdrawn in the 1980s, if not before, as a result of growing international debt problems, leading to the imposition by the IMF and World Bank of structural adjustment programmes. As a consequence, urban-negotiated food price mechanisms and imported food donations seriously undermined peasant enthusiasm and initiative to increase productivity.

Population growth dynamics, land availability, soil quality, climatic factors, transport and marketing systems etc. can play havoc with any agricultural development theory. Densely peopled areas with high natural growth rates impose ever-increasing pressures on land resources, leading to problems of fragmentation, landlessness, soil degradation (especially where lack of credit makes the purchase of nutrients impossible and farmers lack access to expert advice on soil erosion prevention techniques), encroachment on environmentally fragile, vulnerable land and acute pressure on limited water resources. Elsewhere, in rural communities accessible to towns and cities, mass rural-urban migration may have an extremely debilitating effect on agricultural production and rural life in general. The younger generation is drawn away from the land, leaving behind enfeebled elderly men and women to carry out arduous, back-breaking field tasks. In some cases animal herding has to be abandoned because child herders are no longer readily available and elderly, experienced herders are diverted to field tasks because of a desperate labour shortage.

One of the most interesting theories to emerge in recent years focuses attention on the success of the 'non-reform', rather than the 'reform' sector. According to Thiesenhusen (1995, p.7): 'As reform proceeds, some owners in the nonreform sector hesitate to invest for fear of losing their venture capital; but if reform takes place as unused land is expropriated, some owners will press idle property into use in order to keep it'. Alexandratos (1995, p.322) refers to the role new technology has played in weakening the inverse relationship between yields and farm size. He observes that the threat of the reform may encourage owners of large estates to promote modernization as defensive action: 'In some Latin American countries the threat of expropriation and incentive policies (input subsidies, tax breaks) were successful in inducing large farmers to modernize hence increasing outputs ... to render expropriation with compensation very costly'. He cites the example of Colombia, where 'larger farmers often successfully used their influence to extract promises from the government that their land would not be expropriated if they modernized ... as a consequence redistribution of land to the poor was negligible'.

It is clearly apparent that, although the basic aims of agrarian reform are economic efficiency and social justice, political issues intervene at all levels of decision making, planning and implementation. This is inevitable in a situation where 'property owners usually predominate in the structures of political power in most countries, and are the least likely class to change one of the underlying bases of their own status.. For these reasons ... land reform has always proved an extremely difficult proposition' (Ellis, 1993, p.160).

Since the publication in 1967 of Gunder Frank's *Capitalism and Underdevelopment in Latin America*, numerous development economists have theorized about the exploitation of rural areas in Latin America by urban centres and capitalist forces: 'internal domination' and 'internal colonialism' are terms frequently applied to the struggle for land. Dorner (1992, pp.17-18) highlights a new element in the power structure, counterbalancing urban middle class support for the desperate-for-land peasantry: 'Traditional landowners have joined the modern, commercial and industrial entrepreneurs to form the present ruling classes', making the 'political prospects for reform more remote'. The 1991 FAO *Report on Progress under the Programme of Action of the World Conference on Agrarian Reform and Rural Development* concluded that progress in redistributing land 'has been limited mainly because the implementation of land distribution was strongly affected by political realities' (Alexandratos, 1995, p.320).

As to be expected, since the heyday of Latin American agrarian reform legislation in the 1960s, land and development economists have also expended

11

considerable energy and time on theorizing about the aims, requirements, conditions etc. of agrarian reform programmes. For example, Jacques Chonchol, one of the advisers on Chile's original agrarian reform policies, identified in 1964 'eight fundamental conditions of agrarian reform in Latin America: the first and last are particularly interesting. The first condition recognized the importance of water resources: 'Agrarian reform must be a massive, rapid and drastic process of redistribution of rights over land and water'. The eighth condition stressed the need for agrarian reform not to be implemented in isolation but to be 'integrated with a general development plan of the agricultural sector'. In the same year the Inter-American Economic and Social Council enumerated the structural changes required to guarantee the effectiveness and long-term success of agrarian reforms.

Agrarian reform must:

a Produce a change in the structure of land tenure that will permit the increasing of the income of the farmer and achieve the best possible combination of factors for agricultural production.

b Give the land its social function, preventing land and the income it generates from becoming instruments of speculation and economic domination.

c Modernize rural life by incorporating the farmer into the national economy and facilitating the growth of demand for products of other sectors.

d Improve the power structure by giving the farmer a real voice in decisions and a share in political, economic and social opportunities (Stavenhagen, 1970, p.99).

Fine sentiments, but, unfortunately, the Council was obliged to admit that 'efforts to date have been insufficient to achieve the objectives of agrarian reform'.

Types of agrarian reform

Sufficient time has now elapsed to enable researchers to make attempts to classify strategies for rural and agricultural development and to categorize Latin American land and agrarian reforms. Thus Abbott and Makeham (1990) identify three strategies for rural development: modernization, reform and 'deep structural change'. They maintain that, apart from Cuba, all Latin American reforms fall into either the modernization category (in which 'technologies and entrepreneurial methods developed elsewhere are adapted to local conditions without specific reforms in social structure'), or the reformist category ('based on legislative action to eliminate large scale landlord control by allocating holdings above a specific size to former tenants and landless people' i.e. by adopting land reform measures).

Thiesenhusen (1995, p.27) acknowledges the diversity of reform policies: 'The ways in which the Latin American countries have approached land reform are quite different, ranging from the revolutionary situations of Mexico, Bolivia, Nicaragua and Guatemala to the legislated reforms of Chile and the quasi-military authoritarian reforms of El Salvador'. He also draws a distinction between 'minimalist' and 'major' agrarian reforms. Minimalist reforms he defines as 'limited efforts' in response to political pressures (from militant *campesinos* and 'socially conscious dissidents') and with the objective of 'obtaining foreign funding'. Thiesenhusen identifies three major reform types in Latin America: (1) 'reforms that opened new development paths for agriculture and are today considered potential candidates for radical change' e.g. Mexico and Bolivia (anticipating the latter's *Ley INRA* of 1996); (2) 'failed reforms that were enacted by one government to mitigate the social ills of the peasantry only to be rolled back by a subsequent administration' e.g. Chile and Guatemala, though the former also has some characteristics of the first type and (3) 'incomplete reforms (of the 1980s) that were born of sharp rural social inequities but were stifled late in civil wars'.

Agrarian reform strategies and policies are moulded by a combination of social, economic, political and geographical factors and events. Latin American agrarian reform laws have inevitably reflected the prevailing theories and priorities of the decades in which they were enacted. In the 1960s and early 1970s land colonization schemes and worker cooperation featured prominently in agrarian reform planning. By the 1980s 'the huge and profitable plantations, ranches and estates', producing for export and for the rapidly expanding cities of Latin America, were being seen by governments as 'fundamental to the ability of Latin American countries to survive and prosper in the modern world economy'(Gelber, 1992, p.145). In the 1990s, and especially since the 1992 Rio 'Earth Summit', there has been a growing emphasis on the protection of the environment and resource conservation. Today a number of Latin American republics make boast of a Ministry of Sustainable Development: in some, land can be expropriated for environmental reasons, such as protection of biodiversity, without any recourse to land or agrarian reform legislation.

Implementation processes have varied from country to country. 'In countries like Chile, El Salvador, and Nicaragua, insignificant reforms were preludes to more ambitious programs. But in Venezuela, Colombia, Brazil and Paraguay, minimalist reforms were ends in themselves and did not lead to more fundamental alterations of agrarian structure' (Thiesenhusen, 1995, p.162). In some Latin American republics agrarian reform has been phased i.e. a number of laws were passed focusing on different aspects of reform. For example, Ecuador, which 'escaped the political extremes of other Latin American countries' (Gelber, 1992, p.236), introduced four reform bills between 1964 and 1977. The first was primarily concerned with allocating *hacienda* land to Andean *campesinos*; a reform of 1970 expropriated land in the lowlands and encouraged the beneficiaries to form worker cooperatives; a third in 1973 was aimed at raising agricultural productivity and the final one promoted colonization in Amazonas.

Perhaps not surprisingly, some development economists maintain that agrarian reform has run its course and is no longer relevant in an age when market forces are 'modernizing' agriculture throughout Latin America. Such viewpoints are certainly not shared by the millions of land-hungry peasants and large sections of some urban populations. A Brazilian poll in January 1997 established that 86 per cent of those interviewed favoured some type of agrarian reform. Nor do all agrarian analysts agree that agrarian reform has had its day, as the following comment by Zoomers (1997, p.59) illustrates:

> After a decade of discussions about macroeconomic reform and structural adjustment in many Latin American countries, the issue of land reform is back on the political agenda... Comparing the present debate with earlier discussions about land reform in the 1950s and 1960s or agricultural colonisation in the 1970s, it is striking that the concepts and assumptions are still the same, although the context has completely changed. The Latin American countryside is now characterized by greater complexity and diversity than before.

It seems appropriate to end this chapter with a quotation from *World Agriculture: Towards 2010*, in which Alexandratos (1995) summarizes the FAO's current thinking on land and agrarian reform:

> Land reform will continue to be a relevant issue in the future in the quest for poverty alleviation and more equity in the rural areas. However it may cease to be the burning issue it once was, especially in those countries where the non-agricultural sector will be increasingly the main source of additional employment and income-earning opportunities, and land will lose its primacy as the main source of wealth.

2 Agricultural systems in pre-Columbian Bolivia

Bolivia's Altiplano (*altiplano*, high plain) covers slightly less than 10 per cent of the country's surface area, yet from time immemorial it has exerted an unchallenged, dominating influence in terms of population, settlement, agriculture and mining. Built largely of sedimentary, water-laid deposits during the Upper Pliocene, its altitude varies between 3,075m and 4,310m above sea level. From the Peruvian border just north of Lake Titicaca, the Altiplano stretches some 840 km southwards to Bolivia's border with Argentina and has an average width of 140 km. A classic example of an intermontane plateau, it is enclosed by two towering Andean mountain chains. The Cordillera Occidental (the Western mountain range) has been described as 'an accumulation of volcanic material studded with occasional isolated or grouped volcanic cones' (Carter, 1971a, p.7); the plateau is flanked on its eastern side by the imposing Palaeozoic, mineral-rich massifs of the Cordillera Real (the Royal mountain range).

Osborne (1964, p.9) aptly described the Altiplano as 'a high, bleak, barren wind-swept tundra with the paradox of a tropical sun blazing through frigid air'. At first sight it is difficult to imagine why ancient civilizations should have been so attracted to such an inhospitable environment, with seemingly insurmountable barriers to settlement and cultivation: semi-aridity, by reason of being cut off from rain-bearing winds; steep slopes with dry soils, often very sandy or composed of loose, volcanic ash; strong winds exacerbating soil erosion; great diurnal temperature ranges; lack of easy communication through the mountains and, of course, the ever-present threat in some localities of volcanic activity. Early settlers viewed the landscape very differently. The Andean region provided excellent defensive sites; some volcanic soils were extremely fertile for cultivation purposes; hillsides could be terraced and the tremendous altitudinal range made it practicable to grow an equally wide range of crops up to at least 3,700 m.

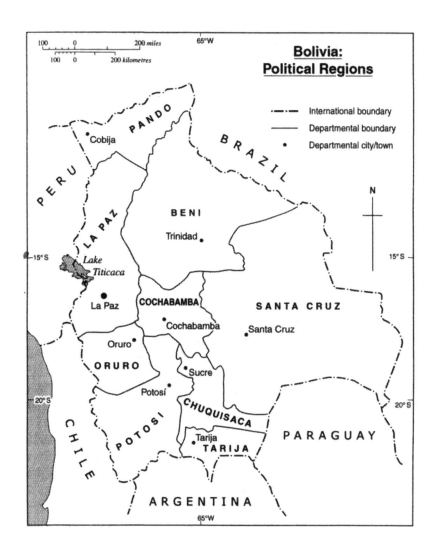

Bolivia: Political Regions

16

Highland dwellers were spared the insect-borne diseases and the fearsome animals roaming the expansive, frequently flooded Oriente (eastern) lowlands, which were, in any case, largely inaccessible.

Archaeological evidence suggests that human occupation of the Andean highlands began over 20,000 years ago. From about 10000 until 8500 BC extensive areas of the Bolivian Altiplano were inundated by glacial floodwaters: the level of Lake Titicaca is believed to have been five metres higher than at the present time (Bouysse-Cassagne, 1991, p.486). Over the next 5,500 years much drier conditions prevailed. In the more sheltered locations on the Altiplano hunters and gatherers of the Viscachani civilization began to domesticate native highland plants and Andean camelids. Archaeological surveys have revealed that by 2500 BC community life and sedentary farming had become well established; a number of Altiplano settlement sites have been discovered on ancient river terraces. Various cultures emerged over the following centuries. For example, Wankarani cultivators and herdsmen occupied the area north-east of Lake Poopó from about 1200 BC until the third century AD. In the Lake Titicaca region the Chiripa culture (ca.1800-200 BC) gave way to the much better known - and still, intensively studied - Tiwanaku civilization. Whilst some pottery artefacts discovered at the main Tiwanaku site have been found to date back to approximately 100 AD, recent archaeological excavations indicate that between 700 and 1300 AD Tiwanaku developed into an extremely important pre-industrial city, with a population probably exceeding 100,000. Although the invading Incas were to discover a ruined city, occupied by Aymara Indians unaware of its origins or the reasons for its catastrophic collapse, it is now believed that at the zenith of its power, both as a trading and ceremonial centre, Tiwanaku dominated an extensive highland kingdom.

A number of today's Aymara farming communities clustered around the Bolivian shores of Lake Titicaca and strewn across the Altiplano claim to be able to trace their ancestry back to pre-Columbian times. Although the Inca people left an indelible mark on Andean landscapes, it was the Aymara Indians who held a commanding position in the highlands from the end of the twelfth century AD until the Spaniards arrived on the scene. It is widely accepted that by the fifteenth century at least seven sizeable Aymara kingdoms had become well established on the high plateau of what was eventually to become Bolivia and southern Peru. Two of the most powerful nations, Lupacas and Collas, together with Pacajes, had built nucleated settlements along the shores of 'the sacred lake', 'the cradle' or 'heartland' of the Aymara peoples. Before the Inca warriors reached the area, some of the kingdoms had even established gold and silver mining colonies 'making the Aymara the premier gold producers of the Andes as well as the leading herdsmen'. So rich were some of these kingdoms that 'they were still considered to be unusually wealthy provinces in the sixteenth and seventeenth centuries' (Klein, 1992, p.18).

Whilst the warlike Aymara were able to subjugate the Puquina-speaking Urus, living in sporadic settlements across the Altiplano and in the Lake Titicaca region, they eventually succumbed to the overwhelming pressures of Inca forces in the 1460s. By the end of the decade a number of Aymara kingdoms, weakened by internal feuds, had lost their independence; as a result of a major revolt against the Incas in 1470, the remaining lake kingdoms were subdued and became incorporated in the Inca province of Kollasuyo.

'It is hard to say just what effect the Inca occupation had on the people of highland Bolivia' (Carter, 1971a, p.30). Some of the more troublesome Aymara groups might have been sent to work in other parts of the empire: the *conquistadores* (invading Spaniards) were to find Aymara-speaking people living as far north as present-day Ecuador. Likewise, some of the Quechua-speaking Inca families dispatched from Cuzco to colonize land in Kollasuyo appear to have remained there permanently. Aymara families living on the Altiplano and around Lake Titicaca had to pay tribute to their overlords: this was exacted in the form of food, produced under *mita* (the Incas' drafted labour system) conditions. It is assumed that, apart from this, Aymara families were left largely to their own devices i.e. that life continued much as before, under the same leadership structures. On occasions when Aymara men were required to take part in empire-building military campaigns, they fought under their own leaders. As Klein (1992, p.20) observes: 'That the Aymara of Kollasuyo retained their languages and autonomous social, economic and even political structures to such an extent is a tribute to their wealth and power in the pre-Incaic times as well as their sense of powerful ethnic identity'. Exactly how many people were living in the Andean region at the time of the Spanish Conquest we will never know. Wearne (1996, p.10) claims that 'the population of the former Inka Empire at the time of the conquest, anything between 9 and 18 million according to the latest research, had fallen to 1.3 million by 1570 and just 600,000 by 1630'.

Whereas Inca warriors were able to impose their authority with ease on groups of non-Aymara-speaking people living in the Yungas (the semi-tropical, deeply incised valleys of the Cordillera Real) and the more easterly *sub-puna* valleys, they met with little success in situations where groups were nomadic. Thus they were unable to extend their control over the tribal groups of hunters, gatherers and transient village dwellers scattered across the vast eastern lowland plains. Such groups, later collectively referred to as Chiriguanos, varied remarkably in size and 'level of development': Siriono-type hunters and gatherers existed alongside 'the sophisticated dwellers of the Mojos floodlands', people who 'were major causeway builders' and 'maintained a year-round settled agriculture' (Klein, 1992, p.23).

Many of these lowland groups remained excluded from mainstream Bolivian economic and political life until the 1990 march for 'Territory and Dignity' publicized their struggle for titled rights to their traditional lands. A 1995 Census of lowland indigenous people established that there are '35 original ethnic groups with a population of 155,118: the size of the groups ranges from that of the Chiquitana (47,000) and Guaraní (36,000) to those of the Leco (9) and Maropa (12), groups sadly on the verge of extinction' (*Bolivia Bulletin*, 1996). In present-day Bolivia the term *pueblos indígenas* normally refers exclusively to these lowland ethnic groups i.e. it is not applied to Aymara- and Quechua-speaking Indians, together comprising more than two-thirds of the republic's population.

Pachamama

Traditional Aymara and Inca/Quechua land tenure and agricultural systems can not be fully appreciated without an understanding of deeply engrained feelings and beliefs about the land and earth. Throughout the ages, *Pachamama* (Mother Earth) has been respected and venerated by Aymara and Quechua people alike, as both a goddess and a mother. Calixto Quispe, an Aymara, explains the age-old relationship between *Pachamama* and the Aymara thus:

> The earth is the source of life, where we are born and where we die. We don't want to abandon the earth although, at times, it seems to abandon us ... The earth for the Aymara is ... a mother who nurses us and from whom we eat. Were it not for the earth, we would not live. Thus, the earth has a power, a religious power and also a tremendous social and cultural power. It is our mother and we call it *Pachamama* (*LP*, 1988).

Aruquipa (1997, *Bolivian Times*) refers to some of the rituals traditionally associated with *Pachamama*: 'The *Pachamama* ... is worshipped in almost all situations Aymaras deal with everyday. Activities such as building houses .. planting crops or even holding community meetings are to be done only after thanks are given to *Pachamama* by offering a *chálla* (the spilling of alcohol on the earth)'.

During the month of August i.e. after harvest and before planting time, llamas have always been sacrificed to succour a hungry *Pachamama*. When the Aymara celebrate new year on 21st June - the year 5506 began for the Aymara on 21 June 1998 - ritual offerings are made to *Pachamama*, to *Tata Inti* (Father Sun) and to various other gods, thanking them for their generosity during the course of the previous year and pleading for good fortune throughout the ensuing year.

As Wearne (1996, p.23) affirms: 'Mother Earth ... has to be apologized to before being tilled, thanked for its harvest and nourished with sacrificial offerings'. The Aymara retain their beliefs that 'unnatural acts' can offend *Pachamama*, who must be appeased. Thus crop disasters (caused by floods, drought, storms, hail etc.) or animal deaths are seen as manifestations of *Pachamama's* displeasure, because an individual or group has committed some misdemeanour; the miscreants must be found and punished to restore the balance of nature. In Aymara communities the field work of a pregnant woman has always been highly valued because through it she conveys her fertility to the soil, whereas menstruating women, symbolizing infertility, are discouraged, if not forbidden, from carrying out field tasks.

Most important in terms of tenure systems in pre-Columbian Bolivia was the strong belief shared by Aymara and Inca Indians that, because *Pachamama* is a goddess, she, i.e. the land, could not be owned. 'Enclosing, owning and selling land, the concept of private property, was a largely alien concept to all indigenous peoples of the Americas. They had a sense of territory, but it was seen as communal and inalienable; land to them was on a renewable lease from a higher authority' (Wearne, 1996, p.104).

Aymara and Inca landholding structures and agricultural systems

In both Aymara and Inca pre-Conquest societies landholding structures were based on the *ayllu*, a clan or kinship system, with close social ties and collective 'ownership' of land. According to McBride (1921, p. 4), 'this common possession of the soil ... seems ... to have dated from the very beginnings of Aymara culture and to have been the foundation upon which the social and political, as well as the agricultural, organization was built, both among the Quechuas and the Aymara'. Bollinger (1993, p.67) refers to the difficulties experienced by the Spaniards in understanding the concept of the *ayllu* but affirms that in pre-Columbian times it was essentially a social structure. Although the *ayllu* may well have originated as an exclusively social organization, it 'took on an agrarian character as the people became more sedentary in their life, the land replacing the family as the bond of union' (McBride, 1921, p.5). Several *ayllus*, drawn together by holding their land in common, usually co-existed inside different communities. Each *ayllu* was split into two distinctive social groups viz. the *hanansaya* or *aransaya* (the upper class division) and the *urinsaya* (the lower class).

Aymara *ayllus* divided their land into cultivable and grazing categories. All families were guaranteed access to pasture land, generally the poorer, more rugged areas, on which *guacchallama* (flocks of llamas and alpacas) were herded by members of the clan designated to perform this task. Arable land was distributed amongst *ayllu* families on an annual basis. Normally each head of household was allocated a *sayaña* (or *tupu*), about a hectare in size. Individual *sayañas* were fragmented into several, sometimes widely separated small plots, with different types and qualities of soil; this enabled a range of crops to be grown,

20

simultaneously ensuring that no *ayllu* members were either unfairly favoured or disadvantaged. A *sayaña* of the same size was allocated to each son; if sons married within the clan they were allowed to retain their entitlement to annually distributed land. Daughters each received a half *sayaña* and lost their access to *ayllu* land on marriage. The Aymara were well aware of the need to conserve soil fertility and ensured that fallow periods were strictly observed.

The harsh climate of the Altiplano, prohibiting the cultivation of tropical and semi-tropical crops, induced farmers to adopt alternative strategies. Thus, by the time of the Spanish Conquest, an intricate network of inter-regional transactions had become well established. Groups of colonists and *kurakas* (regional chiefs) were farming in the Yungas, producing fruits, maize and coca; others were working on the land in the *sub-puna* valleys of the Cochabamba region. Itinerant colonists traded in fish and salt from the Pacific coast. Additionally, landless Urus exchanged fish from Lake Titicaca for root crops and cereals grown on the Altiplano and worked as labourers for Aymara *ayllus*. Murra (1975) has appropriately described this system of exchanging products from widely diverse environments as 'vertical ecological integration'. Whereas today an amazing assortment of agricultural products are bartered and sold in weekly markets on the Altiplano, by producers and entrepreneurs from all parts of Bolivia and beyond, 500 years ago this function was being carried out by the *mitimaq* (the Aymara highland colonists).

According to Von Hagen (1957, p.64), 'agriculture was the soul of the Inca Empire; it determined everything'. Although land was controlled by the *ayllu*, it was regarded as the property of the state i.e. the emperor held the land in trust for the people. State ownership applied equally to most of the llama herds, in addition to the gold and silver mines. Each autumn *ayllu* land was split into three unequal parts: the produce of the first was destined for the Inca (emperor) and the state, of the second, for the priests and religious ceremonies (including the worship of *Inti*) and the third for *ayllu* members. The actual dimensions of non-communal fields were determined by the number and size of families in the *ayllu*, since families were allotted what was considered to be sufficient land to enable them to be adequately nourished. Communal land was reallocated each year in accordance with the changing needs of individual families. As in Aymara *ayllus*, sons were allocated as much land as their parents, whilst daughters received a half measure. 'The boundaries of the fields, especially those separating communal, state and church properties, were well-marked, and their removal was a great and almost unheard-of crime' (Mason, 1957, p.177) - a far cry from the situation today!

The 'sacred fields' had to be cultivated first: even the emperor was obliged to make a gesture, by breaking the first clod with a gold digging stick. Although the field tasks were reputedly accomplished *en masse*, in fact each family was assigned a specific plot to cultivate. Once the work was completed, the same task had to be performed in government fields before the *ayllu* members were allowed to tend to their own plots. After harvest, agricultural produce from the state and 'religious' fields was stored in separate *tambos* (warehouses). Produce from state warehouses sustained the nobility, state officials, soldiers, craftsmen, the aged, widows and the infirm: food was also held in readiness for times of need i.e. famine and other calamities. Highland pasture lands, often vegetated by coarse, tussocky *ichu* grass, were customarily split into the same three categories. *Ayllu* members were only allowed to keep 10 animals. Wool from the government herds of llamas and alpacas was gathered into *tambos* and allocated to families according to their size, to enable them to make clothing for all their members.

Although theoretically in both Aymara and Inca agricultural communities, land was communally held and private ownership prohibited, there were exceptions to the rule. At the local level, 'the plot of ground upon which each house was built seems to have been held almost as private property that descended from generation to generation as a possession of the family' (McBride, 1921, p.6). Within Aymara society, the *kurakas* owned land outside the *ayllu* structure and 'extracted free labor from the *ayllu* members they governed... Thus between the kings, the regional nobles, and the local elders, there existed a group of individuals with access to private property and with inheritable rights to land and labor independent of the *ayllu* structure' (Klein, 1992, p.17). Likewise Inca nobles enjoyed landownership rights and permitted Aymara nobles who offered them no resistance to retain their private properties.

Terracing of steep-sided valleys, which involved enormous expenditure in terms of skill, manpower and time, was considered by the Incas as a vital strategy in order to create additional cultivable land and conserve soils. 'Irrigation ... was the lifeblood of empire ... it was not an Inca invention but it was an Inca perfection' (Von Hagen, 1957, p.66). Inca imprints on the landscape of the Lake Titicaca region will be discussed later, as will the pre-Aymara - or possibly, early Aymara - *suka kollus* (raised fields) of the Tiwanaku area, which continue to attract considerable interest and amazement at their degree of sophistication. Recent experiments have clearly demonstrated that, by applying ancient *suka kollu* farming techniques, designed to protect crops from penetrating frosts, to provide them with a ready supply of water and maintain soil fertility, crop yields can be increased by at least 700 per cent.

The Spanish Conquest

The Treaty of Tordesillas (1494), endorsed by Pope Alexander VI, divided the 'undiscovered world' between the rival maritime powers, Spain and Portugal; it also justified the rights of the *conquistadores* to exploit newly-discovered territory and mineral resources. 'Theologically it was declared that God had granted these lands to the Spaniards as a providence or reward for their wars against the infidels during the Reconquest of Spain ... From the basis of political philosophy it was asserted that there were no legitimate owners of these lands and therefore the Europeans could legitimately conquer them' (Sobrino, 1992).

A later papal edict, *Sublimus Deus*, was rejected by the conquerors: 'Basing their argument on Aristotle's concept of nature, they claimed that the Indians were predestined by nature to servitude' (*LP*, 1991). In 1537 the Vatican proclaimed that 'the Indians were persons with soul and reason, but it blessed crime and pillage: the Indians were possessed by the devil and therefore without rights' (Galeano, 1992, p.9). The conquering Iberians were thus fully sanctioned to 'claim that the Americas were legally *terra nullus* or *vacuum domicilium*, an empty land, unclaimed and unsubdued' (Wearne, 1996, p.104) and to force the occupants of newly-acquired territories into a life of labour and servitude on the land or in the mines. According to the theologian, Bartolomé de las Casas, who attempted to defend Indian rights for over half a century, the *conquistadores* treated the Indians 'not as beasts, for beasts are treated properly at times, but like the excrement in a public square' (Koning, 1991, p.70).

In 1538, six years after embarking on his conquest of Peru, Francisco Pizarro led some of his troops south-east to the Lake Titicaca region in order to subdue a combined force of rebel Inca and Aymara Indians, the latter from the kingdom of Lupacas, which bordered the south-western shores of the lake. After an overwhelming victory, 'given their consistent superiority in armor, steel weapons and horses' (Klein, 1992, p.33), Pizarro returned to Cuzco, the former capital of the Inca Empire. He left behind his two brothers, Hernando and Gonzalo, 'to undertake the full colonization' of what are now referred to as 'the Bolivian highlands and valleys' i.e. the region which would be known by the Spanish colonizers as Charcas and Upper Peru. Some months later this expeditionary force founded the city of La Plata, eventually to become Sucre in 1825.

Once challenged by a priest in Peru to set a Christian example, Francisco Pizarro is alleged to have replied: 'I have not come here for such reasons. I have come to take away their gold' (Blakemore, 1966, p.36). In 1545 his wildest dreams were realized (although he himself had been assassinated four years earlier) when the Spaniards discovered the incredibly rich silver veins of Cerro Rico. Silver mining was to enable Potosí located at the foot of 'the rich hill' (and at an altitude of 4,170m) to become the largest city of the Americas by the early seventeenth century and, reputedly, the world's richest city. As the *South American Handbook* (1998, p.319) informs travellers: '*éste es un Potosí*' is a phrase still used in Spain 'for anything superlatively rich'. The city of La Paz, originally named The City of Our Lady of Peace, was founded by Alonso de Mendoza in 1548, at what is now the small Altiplano town of Laja. Soon after it was moved to its present site, in a gold-bearing canyon protected from the bitter winds of the high plateau and on the main trade route between Potosí and Lima. La Paz rapidly became a thriving centre of commerce, especially as much of the

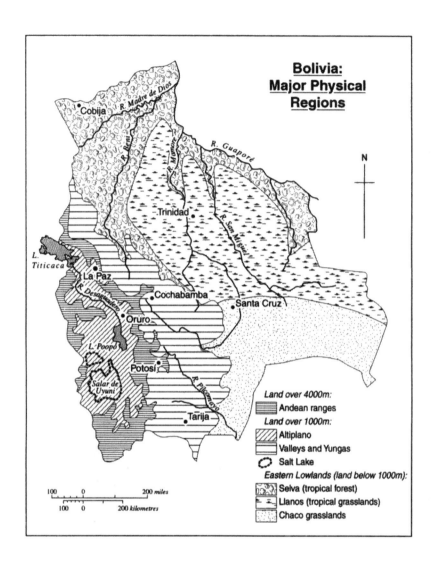

**Bolivia:
Major Physical
Regions**

N

Cobija

R. Madre de Dios

R. Beni

R. Mamoré

R. Guaporé

Trinidad

R. San Miguel

L. Titicaca

La Paz

Desaguadero

Cochabamba

Oruro

Santa Cruz

L. Poopó

Potosí

Salar de Uyuni

R. Pilcomayo

Tarija

Land over 4000m:
Andean ranges
Land over 1000m:
Altiplano
Valleys and Yungas
Salt Lake
Eastern Lowlands (land below 1000m):
Selva (tropical forest)
Llanos (tropical grasslands)
Chaco grasslands

100   0   200 *miles*
100   0   200 *kilometres*

silver mined in Potosí passed through the city on the long, tedious journey to the Peruvian ports - and impressed labour passed in the opposite direction, bound for the mines.

Precisely what the *conquistadores* thought of the Indians living on the Bolivian Altiplano is open to speculation. Some of the Spaniards had no doubt been astonished by the contents of the *tambos* they had raided in Peru. Von Hagen (1957, p.65) refers to Francisco de Xerez, the first soldier-chronicler of the Conquest, who remembered these storehouses as being 'piled to the roof, as the Merchants of Flanders and Medina make them'.

For the most part the *conquistadores* encountered Aymara-speaking Indians, inhabiting rough, *adobe* dwellings, thatched with *ichu* grass, or with *totora* reeds gathered from the shores of Lake Titicaca. Families were cultivating strange roots, such as *papas* (potatoes), *ulluku* (now usually referred to by the Spanish name, *papalisa*), and *oca*: *papas* and *oca* they consumed in dehydrated form. Instead of growing wheat, barley and other European cereals, they produced crops of native cereals, such as *quinua* and *cañahua*. They had no knowledge of cows, sheep, donkeys or pigs (except the native guinea pig); they had never before seen horses. Camelids, native to the Andean region - alien-looking creatures - had been domesticated and lived at altitudes of up to 4,600 m. Llamas were prized as beasts of burden, for their sun-dried meat, *charqui* (from which our term 'jerked meat' is derived), as a source of dung and were also used sacrificially; alpacas produced a wool finer than any European sheep wool, though not as fine as that of the Andean wild, deer-like *vicuña*. To work the land the Indians used a simple digging stick and a primitive foot plough (*taclla*), sometimes bronze-tipped: the wheel was a novelty to them. The Indians liberally imbibed a potent beer, *chicha*, which they brewed from maize. To ward off the cold and hunger, they incessantly chewed coca to which lime-ash had been added.

Certainly the Altiplano must have appeared as an exceedingly inhospitable region for settlement and farming. The *conquistadores* may well have had similar views to those of Osborne more than 400 years later (1964): 'The Altiplano ... is a harsh inhuman land, inhabited by a dour, unsmiling, toiling race'. Taking all things into consideration, it is more than likely that the Indians were regarded by the Spaniards as 'backward', 'primitive', miserable, godless peasants, living in abject poverty, subsisting on the meagre products of a very curious and inefficient farming system. Such perceptions of pre-Columbian Indian cultures have by no means disappeared. Rodrigo Montoya, a Peruvian anthropologist, observes that for today's 'Hispanists... our history began in 1532. Whatever occurred before then is perceived as a sort of shame which our countries carry like a heavy burden, and which impedes "development" and "civilisation"' (*LP*, 1990). Today it seems almost incredible that even in the 1940s the text book used by students following a 'course in the Geography of South America given at the University of Wisconsin' (later renowned for its Latin American agricultural research) could have contained such disparaging remarks as the following:

> The [Bolivian Indians] constitute an inert, unresponsive mass, unwilling to be instructed, and employing only primitive methods, they are incapable of producing enough food for the nation. The situation presents a strange anomaly: agriculture wholly dependent upon the Indians' labor, yet these same Indians constituting a stone wall against any improvement, and the nation faced with a food-production problem that the efforts of the next hundred years will not solve (Whitbeck and Williams, 1940, p.162).

It is abundantly clear that such denigrating opinions and accusations, especially those levelled at pre-Columbian farming, could not be further from the truth. Thankfully, as a result of the process of 'conscientization', the 'popular education' work of some NGOs, political organization and the protest reactions against the 1992 'celebrations' of 'the discovery of America' and '500 years of conquest', indigenous peoples are themselves beginning to re-evaluate their cultures and traditions positively. As Wearne (1996, p.2) comments: 'Ask those leading' (various forms of protest) 'what one scholar, Javier Albó, has labelled the 'return of the Indian' what they are doing and you receive only slight variations on a theme: 'reclaiming our identity', 'demanding our rights', 'regaining our pride and self-worth', 'recovering our spirit''. In a number of Latin American republics no longer are indigenous people, after centuries of having their cultures totally reviled and despised by dominant sections of the population, ashamed to acknowledge the achievements of their ancestors. As Montoya (1990) proclaims:

> Ours was not an empty continent. Great advances had been made in agriculture, water systems and earthquake resistance construction, for example. Here, the Inca overcame hunger, at the same historical time when hundreds of thousands of people died from starvation in Europe. All we have to do is look at the Inca *tambos* - state warehouses - full of foodstuffs, which were surely among the biggest and best surprises encountered by the Spanish when they arrived in America escaping from misery in Europe.

Likewise archaeology and research in general is enabling us to rewrite the history of highland Bolivia and give full recognition to the remarkable accomplishments, in terms of urban organization, building techniques and farming practices, of the Tiwanaku civilization, with its origins on the shores of Lake Titicaca. Coad (1992) reminds us that: 'As the Roman Empire was dying, Tiwanaku was entering the zenith of a civilisation ... covering an area far larger than that of the Romans and based on agricultural techniques more sophisticated than any in Europe for centuries to come'.

26

3 The domination of the *hacienda*

The Colonial period

Distinctions between Inca landlords and Aymara subjects lost their significance as the *conquistadores* established control over the Altiplano. Aymara and Inca/Quechua Indians alike suffered the painful consequences of the Conquest and Spanish colonial policy. Traditional social structures, where not destroyed, remained constantly under threat. Land and liberty were wrested from individual families and *ayllus*. Enforced labour on the land and in the silver mines deprived peasant farmers of both the time and inclination to continue practising the elaborate, environmentally sustainable agricultural techniques of former times. The high incidence of mining accidents and vulnerability to imported diseases, such as smallpox, influenza and measles, decimated the indigenous population in the sixteenth century.

As Wearne (1996, p.101) comments: 'The conquest was rooted in colonization and occupation of territory, even though initially at least the Spanish and Portuguese *conquistadores* of South and Central America came to loot, while their English and French counterparts in the North came to settle'. Unquestionably, the main - for some, the only - attraction of the Andean region was its enormous mineral wealth. In the absence of any Spanish peasant work force, Indian communities were inevitably regarded as vital sources of both food and labour for the mines. All but a few Spaniards were uninterested in procuring land for agricultural purposes since local food supplies satisfied their needs. For this reason, the structure of the traditional *ayllu* was left intact, whilst tribute in the form of food and labour was exacted under the provisions of the *encomienda* (royal land grant) system. *Encomienda* grants, based on groups of communities, were considered by the Spanish crown as 'temporary expedients to ensure the

27

occupation of newly won territory and to reward its conquerors' (Gelber, 1992, p.132). Although the *encomenderos* (landholders) themselves were given no land titles and no legal powers over the Indians under their protection, they maintained the rights 'to receive tribute levied on the Indians ... during their own lives and those of their immediate heirs' (Gelber). According to Klein (1992, p.37), there were in the Charcas region by the 1650s, '82 such *encomienda*s, 21 of which contained over 1,000 Indians each'. By the turn of the century, the *encomienda* had virtually been replaced by the *hacienda*.

By the 1570s the Spanish administrative authorities had become determined to establish more effective control over Indian communities on the Altiplano, especially to isolate them from distant colonies, which could provide a refuge for groups under pressure. Under the supervision of Francisco Toledo, Viceroy of Peru, an ambitious programme was launched to completely restructure Indian settlement. The *reducciones* scheme required Indian families to move into large, nucleated villages, which could be more easily taxed and supervised than scattered communities. Although many newly created villages/small towns were later deserted and some *ayllus* managed, against the odds, to maintain contact with *ayllu* colonists elsewhere, the new *comunidad indígena* (Indian community) gradually took root and acquired a workable form of self-government. A number of today's Altiplano communities are survivals of the Toledo *reducciones*.

Whilst new-style Indian communities were being created, urban settlements were also coming into existence, for the most part as marketing centres for agricultural produce. The current 'boom town' of Santa Cruz de la Sierra was founded in 1561 by Paraguayan-based Spaniards, led by Nuflo de Chávez. It was established as a lowland agricultural post, to supply the rest of Upper Peru with tropical and semi-tropical products, such as fruits, sugar, rice and cotton. Proving too vulnerable to tribal hostilities, Santa Cruz was re-located some 225 km westwards, 50 km east of the Cordillera Real foothills. Cochabamba (originally Villa de Oropeza), recently superseded as Bolivia's second largest city by Santa Cruz, was founded in 1570: it was sited in a broad, fertile valley, later to become one of the most agriculturally productive regions of Bolivia, frequently referred to as the country's 'bread basket'. Not surprisingly in view of its favourable climate and soils, Cochabamba from the start attracted Spanish settlers. The city's fortunes in colonial times were inextricably linked with those of Potosí. In the early seventeenth century it became the main supplier of food to Potosí's flourishing population but its development was thwarted later in the century as the output of silver plummeted.

The vicissitudes of the silver mining industry had an equally strong bearing on the process of *hacienda* expansion. Bolivia's first major phase of *hacienda* growth began in the second half of the sixteenth century and was brought to an abrupt end by the calamitous decline in silver production a century later. The injustices and repression endured by the indigenous population throughout this period are generally considered to have been less severe than those of the second period of expansion i.e. the late nineteenth and early twentieth centuries. Nevertheless, it is clear that many abuses accompanied the attempts to impose a totally alien private landholding structure on a previously Indian-dominated landscape.

A royal decree of 1591, guaranteeing that Indian communities would be allowed to retain sufficient land for subsistence purposes, was largely disregarded. Spanish would-be settlers had little, if any, understanding of communal landholding structures and Andean land practices. Particularly vulnerable to land seizures were the groups of Indians forcibly removed from their original communities to new settlements, leaving behind their unprotected communally-worked farmland. At a time when the Indian population was in decline and peasant resistance at a low ebb, unfenced land could be seized and its expropriation confirmed by Spanish law. '*Denuncia* permitted any Spanish colonist to claim vacant land and, after some formalities and the payment of a fee, to hold it as legal owner. *Composición* permitted him to gain full legal possession of any portion of land under his occupation that suffered from defective registration of its title' (Gelber, 1992, p.134).

In early colonial times some of the more fortunate *ayllus* and communities, not subject to the *hacienda* system, were able to take advantage of expanding marketing networks, particularly in the mining areas, but the vast majority of peasant families on the Altiplano were more exposed to hunger and starvation than in the days preceding the Conquest: whilst Aymara *ayllus* had been subdued by Inca warriors, at least their landlords' warehouses had provided sustenance in situations of acute need e.g. drought and crop failure. In colonial times communities became more vulnerable to such hazards, partly as a result of abandoning traditional practices, such as the maintenance and usage of irrigation channels.

According to Osborne (1964, p.74), 'the Indians were ... driven to subsist at starvation level on the most unfertile tracts of the unfertile plateau or forced to labour in conditions of serfdom on the estates of those who had despoiled them'. Carter (1971a, p.33) quotes an observation made in 1649 by the then Bishop of La Paz about one particular Altiplano province: there were 149 estates 'composed of Spaniards in such a small province that (the Indians found) themselves dispossessed not only of their native lands, but even compelled to secretly hide, preferring the more certain fortune of flight to frequent ruin'. Some *originarios* (original *ayllu* members), who were known by landowners to have escaped from the mines or other obligations, were pressurized by them to renounce their rights to land and work as 'protected' *yanaconas* (landless people) on the *haciendas* (Buechlers, 1971, p.3).

A first agrarian reform?

Unquestionably, Spanish *hacendados* (landlords), whether or not by design, succeeded in transforming the agricultural landscapes of large areas of Bolivia. Some researchers, such as Carter (1964, p.7), have even suggested that the expansion of the large estate system during the sixteenth and seventeenth centuries should be regarded as Bolivia's first attempt at agrarian reform. Although the *hacendados* were certainly not motivated by a desire to extend social justice to the indigenous population, it must be acknowledged that, in most other respects, the massive, aggressive agricultural restructuring 'programme' set in motion by them - and largely approved by the Spanish authorities - does 'fit' the definition of agrarian reform referred to in the first chapter. Spanish estate owners imposed totally alien forms of private landownership and labour; they introduced new farming techniques, such as the use of the common Mediterranean-type plough, cultivated crops unknown to Andean farmers and reared European animals. Whilst the Indians had traditionally practised subsistence farming, Spanish farmers were primarily concerned with selling cash crops in urban markets.

Two-thirds of the Indian population still remained under the effective control of the *hacendados* when the period of *hacienda* expansion ceased in the mid-seventeenth century. At the point when *hacienda* growth was being checked by the decline of silver output, the Indian population was actually increasing, partly as a result of at last acquiring immunity from endemic European diseases. Thus whilst some *hacendados* could not avoid bankruptcy and others struggled to survive, growing 'Toledo' communities were looking for opportunities to expand territorially and even making intermittent attempts to encroach on uncultivated *hacienda* lands.

The crisis in Upper Peru's silver mining industry led to structural changes which, despite the gradual revival in silver production beginning in the 1750s, remained virtually intact until the late nineteenth century. The European, Indian and *mestizo* (the term traditionally applied to a person of racially-mixed blood) populations of Potosí and other mining centres declined dramatically, whilst Cochabamba ceased to function as the mining region's main supplier of food. Meanwhile Altiplano Indians, released for the most part from the burdens of forced labour in the mines, and under less pressure from landlords, contributed in large measure to the sustained growth and prosperity of La Paz as a marketing centre.

The 1780-82 Tupac Amaru-led Great Rebellion, which spread rapidly from Cuzco as far south as northern Argentina and involved the participation of perhaps 100,000 Indian rebels and their *kuraka* leaders, constituted, in reality, an independence movement; it marked the last major concerted attempt by Indian communities and their regional chiefs to regain their freedom by overthrowing Spanish rule. As on former occasions, the Lake Titicaca region featured prominently in the revolt and was the scene of violence and bloodshed. In March 1781 Andrés Amaru (the nephew of Tupac) established control of most of the eastern lake-shore, carrying out considerable destruction to *haciendas* in the process. Subsequently, with the support of Julian Apaza (renamed Tupac Katari), Amaru proceeded to besiege La Paz for several months before being overwhelmed by troop reinforcements from Buenos Aires.

Klein (1992, p. 77) refers to 'the massive loss of farm implements, animals, and workers on abandoned *haciendas*' in the lake region. Whilst most of the *haciendas* were working normally again by the end of the 1780s, the Indian chiefs paid very heavily for daring to challenge Spanish authority. They had done so because of a deep resentment at their loss of privileges and exploitation by *hacendados*: an additional factor for some had been the 'never-ending attack on local religious belief systems'. Once the rebellion had been put down, regional chiefs were forcibly deprived of their lands and removed from office. 'Following the rebellion of 1780-82, the Indian noble class ceased to be a major factor in the social, economic and political life of the region' (Klein, 1992, p.77). Whilst the Indians lost their traditional leaders and sense of direction, the title of *kuraka* was assumed by local Spaniards.

Ecological disaster?

Was the Spanish Conquest directly responsible for 'ecological disaster' throughout the Andean region? Over recent years a number of writers have supported this view. Gelber (1992, p.131) claims that 'the introduction of European livestock had the most devastating impact on the physical and cultural landscape of New Spain'. MacDonald (1992, p.4) highlights the impact of the structural change on traditional agricultural systems in Bolivia and Peru: 'The Spanish defeat of the Inca empire ... tore up by the roots the cooperative agrarian society of the Andean *ayllus*... A way of life that today we would call 'sustainable' was destroyed'. Wearne (1996, p.122) speaks in more general terms about the environmental legacy of the Conquest:

> The environmental and biological consequences of the arrival of the Europeans were in some ways even more devastating than the physical and cultural consequences. Despite the impact of the diseases brought by the Europeans, the flora and fauna which accompanied them may have been even more catastrophic.

The extent to which colonial farming should be blamed for the virtual de-vegetation of large areas of highland Bolivia remains open to speculation: in the absence of detailed research, especially pollen analysis data, opinions vary widely. Mason (1957, p. 137) claimed that: 'Centuries before Inca days the highland plateaux had been practically denuded of the few trees they contained. The farmers of the Inca period gathered all the brush and dead bushes and branches..'. According to Morris (1987, p.13), 'the ease with which blue gum (*Eucalyptus globulus*) and Monterrey pine can now be grown at heights of up to 4,000 m suggest that the valleys between the mountains, from Chile to Colombia, and even the lower *páramos*' (above today's treeline), 'were once covered'. He identifies several culprits: 'Tillage, and cattle farming on steep slopes with dry soils ... tree-felling for construction, fuel or land clearance, have resulted in a tree-less windblown landscape and parched, eroded soils'.

Whether or not they were implicated in tree destruction, sheep have undoubtedly contributed in large measure to today's problems of severe soil erosion and land degradation in many Altiplano communities and in numerous other locations. Prior to the Conquest large flocks of sheep had rendered vulnerable Iberian soils practically useless - a problem which the *conquistadores* and later settlers conveyed to the New World. Of all the animals and plants introduced from Spain, sheep were the most highly prized by Aymara and Quechua farmers. Whilst the rearing of cattle and horses was largely confined in early colonial times to *haciendas*, 'sheep were widely and rapidly accepted by the Indians and even in the 1560s were increasing in numbers in areas such as southern Peru, where there was little Spanish settlement' (Smith, 1971, p.279).

Although the actual initial transportation of sheep to the Altiplano must have posed formidable problems (through disease, water shortage and respiratory difficulties), once there, flocks of sheep were not hard to rear. They provided meat and produced more wool than llamas and alpacas; since they could not survive on the highest *ichu* pasture lands of the *puna*, they posed little threat to native camelids. On the other hand, whilst llamas and alpacas were able to tolerate arid conditions because of their water storage mechanism, sheep competed for limited water resources and land otherwise devoted to high altitude crops, such as *quinua* and *cañahua*. Additionally, whereas llamas nibbled vegetation gently, without damaging the root system of plants, and were able to aerate the soil by reason of their unusual foot structure, sheep tended to tear out grasses and broke up fragile soils, exposing them to the erosive forces of strong winds and occasional storms.

European temperate crops, normally grown in rich, well-watered soils, 'had to find their niche ... in the careful altitudinal zoning of traditional Indian farming' (Smith, 1971, p.279). Of the crops introduced by the early settlers, it was barley, because of its short growing season and ability to yield well up to an altitude of 4000 m, that appealed most to Indian farmers in the Andean region. It could be cultivated alongside native crops of *papas, quinua* and *cañahua*. Wheat, unable to grow at levels above 3,500 m, and thus excluded from all but the most sheltered Altiplano depressions, like maize, proved ideally suited to the climatic and edaphic conditions of the Cochabamba area: in great demand by Spanish settlers for bread making, it was soon being cultivated extensively on large estates.

The Spaniards expressed minimal interest in raising native camelids. On the other hand, they greatly valued the contribution made by sure-footed llamas to the transport of silver along the Silver Road for export from Arica, and also to Lima. With few exceptions, *hacendados* also rejected most indigenous crops as inferior: had they been aware that *oca* is a highly nutritious tuber, and that *quinua* and *cañahua* (bread grains) are reputedly richer in vitamins than any other cereal, it is possible that they might have regarded them more favourably. The only native Andean food plant to be appreciated was the potato, believed to have been cultivated in the Lake Titicaca region since at least 400 BC.

The potato, the second of the Bolivian highlands' two main products, reached Europe some time in the mid-sixteenth century It has been traced in Spain around 1570. Sometimes the upper classes ate it as an exotic curiosity; but by the populace it was usually regarded with profound suspicion. In Ireland, however, it soon became indispensable (Lindqvist, 1974, p.110).

Legislation on land tenure between 1825 and 1952

Bolivian independence was eventually declared in Chuquisaca (to become Sucre) on 6th August 1825, after a prolonged dispute between Simón Bolívar and Antonio José de Sucre over whether or not Upper Peru would be able to have an economically viable existence if detached from the rest of Peru. 'Carved from the Viceroyalty of La Plata and consisting essentially of what had formerly been the *audiencia* of Charcas' (Smith, 1971, p.320), the new republic, named in honour of the Liberator, was originally twice its present-day size.

Bolívar, who for a short time became president of the new republic, was one of only a handful of nineteenth century politicians intent on easing the lot of the long-exploited Indians and integrating them into the regional economy. After the Battle of Ayacucho (1824) had liberated Peru, Bolívar himself had spent some eight months 'travelling through the different provinces of Peru, inquiring into local government, opening schools and trying to improve the conditions of the Indians' (Trend, 1946, p.187). He was convinced that, by individualizing holdings, the Indians could be converted into competitive small farmers and, thus, rural poverty, alleviated. Accordingly, an 1825 decree issued in Trujillo abolished communal property and entitled the owners of holdings to sell or dispose of them at will. Once aware of the immediate and disastrous consequences of his legislation, Bolívar rescinded it and simultaneously prohibited the further sale of indigenous property before 1850. Unfortunately it was already too late to prevent thousands of peasants from entailing or selling their newly created holdings to unscrupulous speculators. A second well-intentioned decree issued by Bolívar, abolishing the burdensome tribute tax, collected from all Indian males between the ages of 18 and 50, was also cancelled within the first year of the republic; tribute tax rapidly became the government's main source of income.

Few other legislative measures impinging on Indian communities were enacted during the early days of the republic. It was, however, a period during which the controversy of the so-called 'Indian question' or 'rural problem' grew increasingly bitter, culminating in the extreme land tenure legislation of President Melgarejo (1864-70). The agrarian question, initially debated academically by the intelligentsia, later divided urban residents and became the burning issue of the day. Luis Antezana (1971) sums up the widely opposed viewpoints of 'the colonialists' and 'the nationalists'. The former, a minority of wealthy landowners and political opportunists, were bent on developing and extending large estates i.e. not averse to usurping communities and subjecting 'free' Indians to the *hacienda's* system of duties and obligations. On the other hand, the majority of mainly *mestizo* and Indian urban dwellers, adhering to '*la nacional*' dogma, were convinced that, apart from economic considerations, any governmental action to promote the landowners' designs for expansion constituted a retrogressive step towards 'feudalism' and colonialism; they insisted that the integrity of Indian communities should be respected and that they should be permitted to develop peacefully and democratically.

'Nationalist', pro-community supporters contended that a number of communities had evolved to such an extent that they could be expected to provide a sound basis for an intensive, progressive system of agriculture, if left to their own devices. It was argued that in certain regions traditional patterns of collective land ownership and of mutual labour had begun to disintegrate and new structures of tenure and production were developing; in a few communities landless *forasteros* ('foreigners') were being paid regular wages by fellow-community members. Bernardino Sanjinés, an avid investigator of the pre-Melgarejo period, claimed that in the department of La Paz, as in the Yungas, community lands were being exploited more efficiently than those of neighbouring *hacendados*: he concluded that in the Yungas the discrepancy in output was in the proportion of four to one in favour of the communities. According to Sanjinés, processes of 'democratic differentiation' were beginning to operate within Indian communities: a number of freeholdings had advanced beyond the subsistence stage and were adequately supplying nearby towns and mines with agricultural produce.

Members of the landholding, 'colonial' group (also opposed to the protection of Bolivian trade and prepared to place control of mineral exploitation in foreign hands) 'justified' their proposed course of action in an attempt to court popular support. José Dorado, one of their most prolific writers, frequently referred to 'the dead hands of the Indian' and upheld the *hacienda* system as the sole effective means of rescuing Bolivia from a position as 'the last of all'. He emphasized the 'benefits' bestowed on the Indian by the *hacienda* system: 'to seize these lands from the hands of the Indian is to convert the poor and miserable peasant farmer into a rich and well-to-do *hacienda* worker' (Antezana, 1971, p.21). It was appreciated by such protagonists that their objectives could only be satisfied after gaining political representation at national level. In the formative years of the republic they rarely encountered success. Certainly, Andrés de Santa Cruz, himself the grandson of an Indian chief from the town of Huarina near the shores of Lake Titicaca, demonstrated little sympathy towards their cause during his decade of office.

As late as 1846 a national census recorded approximately 11,000 communities of which some 4,000 were freeholdings: of the 79,267 families living in communities, 49,293 were said to own plots of land and the remainder to be *forasteros*. The same census registered 5000 *haciendas*, at a time when the *hacienda* was in 'a relative state of stagnation, except for the two exceptional areas of the Yungas and the Cochabamba Valley' (Klein, 1992, p.124).

The landowning influence in government became increasingly powerful and in 1864 estate owners were to find a staunch ally in General Mariano Melgarejo, nowadays generally acknowledged as the most despicable and barbaric of all Bolivian leaders. His presidency heralded a period of 90 years during which virtually all land tenure legislation was destined to extend and consolidate 'the feudal regime' (what Antezana terms 'feudal colonialism'), and attack those who tried to replace it by a democratic system. During Melgarejo's violent years a series of vindictive measures afforded the landowning fraternity their long-awaited opportunity to almost eliminate Indian communities. In the name of the law, *hacendados* purchased communities outright or acquired by force land that since pre-Conquest times had been possessed and worked by Indian families. In 1866 Melgarejo, motivated by an urgent need to replenish an empty treasury, introduced his infamous 'Confiscation decree'. Since, it was stated, all holdings occupied by Indians legally pertained to the state (Bolívar had declared the community dwellers masters of the land they occupied), they were required to pay a tax of not less than 25 pesos and not more than 100 pesos in order to obtain legal titles. Those who failed to do so within the 60 days allotted would be deprived of lands, which could then be auctioned off; if no buyer came forward, the Indians who had previously worked the plots could continue to do so, on the condition that they paid a rental to the state. Few peasant farmers, even had they been made aware of the injunction, could have accumulated the sum in the time allowed and thus more than 100,000 Indians were deprived of their ancestral lands by a single vicious measure. Indeed, once set in motion, sustained by the ruling oligarchy, the alienation of communities proceeded unchecked long after the assassination of Melgarejo. It has been suggested that, as a result of his decrees, in all, more than 600,000 community dwellers were deprived of land over the years. Comparing figures from the 1900 land tax returns with those from the 1854 Indian reviews,

McBride (1921, p.26) concluded that in the intervening period there had been an unprecedented decrease in communal holdings in every province of La Paz, Oruro and Chuquisaca departments and in all but four provinces of Potosí and Cochabamba.

A brief period (1873-95) of reorganization and expansion of Bolivia's silver mining industry, immediately before its dramatic eclipse by tin mining, ushered in the second phase of *hacienda* growth, during the early days of which estate owners were able to take full advantage of Melgarejo's legislation. Whilst the period of land estate expansion is considered by some to have lasted from about 1880 until Bolivia's defeat in the Chaco War, Antezana amongst others claims that, in fact, this period of 'the pillage' of community lands lasted until 1952 i.e. that this prolonged period of mass alienation of the peasantry was only ended by the National Revolution of that year. It was an era during which 'the power of the free Indian communities was definitively broken. Only the marginality of the lands they still retained and the stagnation of the national economy after the 1930s prevented their complete liquidation' (Klein, 1992, p.152).

From the time of Melgarejo, whenever landowners encountered resistance in achieving their expansionist ambitions, deliberate and systematic massacres were organized by regular forces sent into the countryside specifically for this purpose. In 1869, at the Lake Titicaca crossing of San Pedro de Tiquina, 600 *comunarios* (peasants living in free communities) lost their lives in a desperate effort to recover their ancestral holdings from rapacious *hacendados*: shortly afterwards government troops massacred an additional 2,000 peasants in Ancoraimes, another lakeside community. Infrequently peasant resistance made a strong impact nationally. The popular uprising led by Zarate Willka (1898) in the Bolivian highlands, was instrumental in bringing a Liberal government to power. But, in common with a number of agrarian protest movements of the late nineteenth century, Willka's success was short-lived: subsequent regimes ignored former promises of justice for the Indians and Willka himself was assassinated (Huizer, 1973, p.8).

Ruthless usurpation of a *comunidad originaria* (freeholding or free community) normally signified utter destitution for the displaced *comunarios*. A limited number of enterprising community leaders organized rapid retreats *en masse* to rocky, almost inaccessible mountainsides - areas isolated from communication lines, sometimes approaching the very limits of cultivation and thus of minimal value to the landowning class. In such localities *comunarios* managed to eke out a meagre existence under adverse environmental conditions but, at least, free from the victimizations of the *hacienda* system. Amongst such scattered groups, banditry was at times essential to survival and in the late nineteenth century Bolivian prisons were filled with dispossessed *comunarios* and *colonos* (peasants working for estate owners) convicted of pillage and similar offences.

36

Llamacachi, one of the Lake Titicaca communities studied in detail, successfully engaged in a prolonged and fierce struggle as contiguous estates tried to expand at its expense, but such an example, especially in a favourable lakeside setting, is a rarity. For most *comunarios* the only alternative to withdrawal into inhospitable places of shelter was submission to the pressures of the new usurping landlords: whereas a *comunario* might previously have cultivated from 15 to 20 hectares of land and participated in the regional economy, he was now compelled to labour without remuneration for the man who had seized his holding and ruined his livelihood. He was constrained to live at subsistence level, whilst sometimes tracts of land, formerly cultivated to advantage by himself and his fellow-peasants, became superfluous under the new system and were entirely neglected by the *hacendado*. Usufruct rights to such terrain could have been granted without any direct loss to the landlord, but a foreboding that estate workers might at some future date claim squatters' rights precluded their access. The National Agrarian Reform Council later maintained that Melgarejo, by converting *comunarios* into 'feudal *colonos*', dependent for their existence on usufruct rights to minute and scattered parcels of land, gave birth to the abuses of *minifundismo* (fragmentation of plots) - an intractable problem to the present day.

Bolivia's ignominious defeat by Paraguay in the Chaco War (1932-1935) generated widespread disillusion and 'shattered the traditional belief systems' (Klein, 1992, p.187). Yet although the leaders of the National Revolutionary Movement (the MNR), who promulgated Bolivia's Agrarian Reform Decree in 1953, traced the roots of the momentous 1952 National Revolution back to the Chaco War, it is clearly apparent that the seeds of dissension had already been sown in the 1920s - a decade of political crises and violence. For Bolivia the 1920s marked a dramatic turning point socially, politically and economically: it was a period of intense class conflict, radical political change and rebellion, both on the land and in the mines.

Continued *hacienda* expansion on the Altiplano provoked a series of attempted land invasions and peasant revolts throughout the decade. Particularly violent was the major uprising in 1921, in the community of Jesús de Machaca, about 30 km south of Lake Titicaca. In their endeavours to quell the rebellion, President Saavedra' s armed forces resorted to severe repressive tactics against individuals and community authorities; several hundred Indians were killed, houses and crops were burnt and 70 leaders, sent to gaol in La Paz (Dunkerley, 1984, p.24). The early 1920s also witnessed the formation of the country's first labour unions and earliest national strikes, largely attributable to the plummeting output of minerals, as Bolivia suffered the devastating effects of the Great Depression.

It was not until the early 1920s that Marxism made an impact on Bolivian political thought: in 1921 a national Socialist Party was founded. 'A small group of intellectuals ... began to discuss such basic issues as Indian servitude (*pongueaje*), the legal recognition of the Indian community governments, and the rights of labor and of women' (Klein, 1992, p.173). As a wave of liberal ideas infiltrated Bolivian cities, renewed attempts were made to instigate rural reform, partly from humanitarian motives but equally in response to a growing awareness that the *hacienda* system of land tenure and usage was economically inefficient: it was failing to supply agricultural produce in quantities capable of sustaining an expanding urban population and was, on the contrary, acting as a drag on the national economy.

In 1925 a law was enacted prohibiting the sale or disposal of Indian holdings except by auction; instead of protecting peasant farmers, it unintentionally invigorated the process of communal property alienation by openly drawing to the attention of estate owners the fact that land still remained available for purchase. Free communities remained vulnerable, some being incorporated into existing estates which utilized the newly available labour, whilst simultaneously leaving the land intact. In 1929 all estate owners with more than 25 resident Indians in their charge were required by law to construct schools on their properties, but, as President Victor Paz Estenssoro was to remark in an address to the National Congress in 1956, in many cases this did not happen: 'The peasant was better exploited as an ignorant worker than as one who knew his rights'.

By the early 1930s a stagnating national economy had brought to an end the second phase of significant *hacienda* expansion. Over the previous 50 years it is likely that the size of the landless Indian population had doubled: by 1932 there were many more landless peasants than those living in free communities. Whilst an expanding tin mining industry had been largely responsible initially for the transformation of the agricultural landscape and brought untold wealth to the Patiño, Aramayo and Hochschild families, it had made minimal headway in modernizing Bolivian agriculture and for countless communities had spelt disaster.

The Chaco War was instrumental in breaking down traditional rigid class barriers. For the first time Indian foot-soldiers from both farming and mining communities fought alongside white and *mestizo* urban dwellers: Indians returning from the Chaco were not alone in being appalled by the corruption and incompetence demonstrated by the country's political and military leaders. Groups of nationalistic intellectuals intent on social reform were invigorated in post-war years by the support of young ex-officers and other politically aware young people, collectively referred to as 'the Chaco generation'. As a consequence of the war,

Bolivia ... changed from being one of the least mobilized societies in Latin America, in terms of radical ideology and union organization, to one of the most advanced...The result of that change in largely elite thought was the creation of a revolutionary political movement that embraced some of the most radical ideas to emerge on the continent (Klein, 1992, p.187).

38

In the years following the Chaco War, the radical reinterpretation of 'the Indian problem' and 'the land question', propounded by emerging socialist parties and intellectuals, gained widespread credibility amongst urban populations. It was at last acknowledged that the 'backwardness' and 'passivity' attributed to Indians over generations were not the innate characteristics of an inferior race but, rather, defence mechanisms adopted by them when confronting ruthless exploitation by *hacendados*: it followed that the wholesale destruction of the *hacienda* system represented the peasantry's only means of escape from such rank injustices.

In such a political climate it was almost inevitable that, sooner or later, Indian veterans of the Chaco War would challenge repressive landlords, who had previously prevented them from engaging in all forms of political activity. Thus a year after the war ended, Bolivia's first two peasant syndicates were formed at Cliza (Ucureña) and Vacas, in the Cochabamba region, by ex-combatants, local teachers, Cochabamba students and the son of a local landowner. Clark (1970, p.25) identifies the determining locational factors: the area was an important agricultural one, supplying food to the mines; peasants could rely on support from belligerent miners and students; social integration was more advanced than elsewhere in rural Bolivia and, above all, institutions such as the Roman Catholic Church and the city of Cochabamba owned a number of *haciendas*, suitable for the purposes of challenging the system. Accordingly, syndicate members at Ucureña negotiated an agreement with the landowner, the Monastery of Santa Clara, to lease *hacienda* lands for cultivation. Although five local landowners, fearing further anti-landlord agitation from peasants supported by students and miners, retaliated in 1939 by buying the monastery lands, terminating the leases (on the grounds of 'rationalizing production'), evicting the peasants who refused to work for them gratuitously and exiling 12 of the leaders, the experience 'did much to unite and radicalize the Indians, who went on organizing themselves. Ten years later what happened at Cochabamba was repeated on a nation-wide scale' (Lindqvist, 1972, p.114). Ucureña was to acquire greater fame in August 1953 as the location selected symbolically by Paz Estenssoro for signing Bolivia's Agrarian Reform Decree.

The MNR Party was founded in 1942 by a group of liberal intellectuals and army officers, under the leadership of Paz Estenssoro, who in 1943 was taken into the government of Gualberto Villarroel. In the short period before Bolivia's 'years of turmoil' (1946-52), during which the MNR remained illegal, several attempts were made to improve the lot of the peasantry by legal means. Thus in 1944 Paz Estenssoro and Walter Guevara Arce presented a proposal to the National Convention for 'a moderate agrarian programme' (Huizer, 1973, p.51). Whilst opposition from the Society of Landowners was sufficiently strong to block the proposal's acceptance, the moral support given to the peasantry strengthened their resolve to continue their political protests and press for reform.

In 1945 Bolivia's first National Indian Congress, organized by the Indian leader, Luis Ramos Quevedo, attracted some 1,000 delegates, representing all the republic's provinces. Whilst participants refrained from demanding wholesale agrarian reform, they urged the government to abolish unjust *hacienda* personal services and to provide adequate educational facilities in rural areas. In response, President Villarroel issued Decree 319, declaring the hated *pongueaje* illegal and restricting *hacienda* work to four days weekly; additionally, enforced services to cantonal, provincial, judicial and ecclesiastical authorities were abolished (Antezana, 1971, p.147). In July 1946 resentful landlords secured their revenge against Villarroel, perhaps best remembered for his remark, 'I am not an enemy of the rich, but I am more a friend of the poor': he was unceremoniously hanged from a lamp post outside the presidential palace. Subsequently many of the peasant leaders who had attended the Indian Congress were gaoled.

Villarroel's death unleashed widespread peasant protests, which on occasions took the form of land invasions or *huelgas de brazos caídos* (sit-down strikes), but became increasingly violent, sometimes involving the assassination of landlords trying to reintroduce *pongueaje*. In some cases, peasant syndicates, with the help of miners, attacked and occupied provincial capitals: such activity was almost invariably met with brutal punishment administered by a rural police force created specifically for the purpose. Not surprisingly, by 1950 'peasants no longer asked for reform of the *colonato* system ... they wanted its complete abolition' (Thiesenhusen, 1995, p.57).

Contrasting views about Bolivia's landholding structure on the eve of the National Revolution (1952)

The hacendado

The almost sub-human conditions endured by the vast majority of *indios* aroused no feelings of guilt in the typical *hacendado*: to him the estate system was 'economically rational as well as socially desirable' (Lockhart, 1969, p.424). Ownership of land, even though the land in question might have been ruthlessly seized from *comunarios*, was a 'God-given right'. By granting access to cultivable land and discharging their side of the bargain, some landlords insisted that they were actually rendering a valuable service to rural communities by creating self-contained units in which a spirit of paternalism could prevail and protection be afforded against exploitation from outside agencies. Writing in 1967 (p.366), Erasmus records the comments of the wife of a Bolivian *ex-hacendado*: condemning the *lucha de clases* (the class struggle) which she believed had been provoked by agrarian reform, she lamented the fact that the *indio* had lost his former '*cariño al trabajo y al patrón*' (love for work and for the landlord).

From the *hacendado's* personal viewpoint, the institution was ideal. Ownership of land had always been accepted as an indication of wealth and influence and it was usually unnecessary for him to devote much attention to his estate; the *hacienda* provided him with sufficient income to live in relative comfort in La Paz, Sucre or Cochabamba and to educate his children in Europe. Administration could safely be left in the hands of a carefully selected *mayordomo* (*hacienda* administrator). Under the system labour was free and his tenants furnished the animals needed for ploughing and harvesting. Little capital was required to produce crops, meat and dairy products, for which he was assured a ready market. He maintained that if he made any attempt to boost output by 'modernizing', increased production would only 'flood' the market: it was preferable to keep the markets in short supply and, if prices were unfavourable, to store farm produce in the *aljeria* (*hacienda* storehouse/shop in the city) until the time was auspicious for selling it. If he did purchase new properties and leave lands uncultivated, it was often to discourage competition. In short, the *hacienda* was an excellent means of raising one's status and of converting unremunerated labour into saleable commodities.

The colono

An agricultural system bestowing such rich rewards on the landlord, signified oppression and utter despair for the estate worker. In every sense he was exploited by the *hacendado*. As a member of the lowest class of a rigid class structure, he had little possibility of improving his situation; between him and the landed oligarchy was a tremendous gap socially, politically and economically. As Tamayo had remarked in 1910: '*El Indio da todo al Estado pero el Estado no da nada al indio*' (the Indian gives everything to the State, but the State gives nothing to the Indian).

The estate system was to him an institution for exploiting his and his family's labour and offered few, if any, opportunities for securing alternative employment or access to additional land; often, just enough credit would be made available to tie him indefinitely in debt bondage to the estate. Scarcity of land and time normally precluded any notion of production in excess of subsistence needs. If he was more fortunate and able to make outside contacts, it was merely through the sale or barter of small amounts of staple products in local periodic fairs. On occasions it was possible for a *colono* to gain titles through marriage to plots of land in a *comunidad originaria*: in this case he would perhaps be enabled to move temporarily to the house of his parents-in-law to avoid his turn at some of the more exacting personal services, returning to the *hacienda* when there was no longer any danger of recruitment. However, generally marriage was endogamous and a *colono* from outside, considered an unwelcome intruder in the freeholdings. Both *comunarios* and *colonos* belonged to the lowest category of Bolivian society

41

- *la indiada* - but the former, perhaps because they had displayed a 'superiority' by resisting the pressures of the *hacendados*, retained a higher social standing within the group and frequently despised their less-fortunate neighbours. Traditional antagonism dies hard and it was not uncommon, even in the 1970s, for *comunarios* in the Lake Titicaca region to refer arrogantly to *ex-colonos* in the vicinity as '*hombres de llamas*' (men of llamas) or even '*incomprensibles*' (dullards).

Osborne, writing originally in 1954, remarked that 'there is not and never has been on Bolivian soil a large unprivileged class ready and anxious for emancipation to the rights and duties of full citizenship ... in a modern economic state'. It is less condescending and, indeed, more realistic to accept Gunder Frank's assessment of the situation (1971, p.154): 'the self-chosen retreat ... is the Indians' only available means of protection from the ravages and exploitation of the capitalist system'. The *colono* was excluded from access to education and the Spanish language, from political organization and, by inadequate income and literacy qualifications, from national politics. There was no institution outside the estate to which his problems and protests could be borne since local and national governments were content - perhaps relieved - to leave management of tenants almost entirely in the hands of the landowners: moreover, frequently provincial administration was controlled by the *hacendados*. All things considered, the *colono* was a serf and advertisements for property sales made reference to the number of families resident on the estates and therefore available for labour.

The economist

By 1952 Bolivia's *latifundio* landholding structure was 'one of the more extremely concentrated structures in Latin America' (Clark, 1970, p.4). It was an unjust, exploitative and inefficient system, enabling landowners to control access to the republic's best lands. So widespread had been the alienation of Indian freeholdings, that only 3,783 had survived in their traditional form, these being almost exclusively confined to the plateau departments of La Paz, Oruro and Potosí.

Bolivia's first Agricultural Census of 1950 dramatically exposed the gross inequalities in landownership. Of a total 33 million hectares classified as 'cultivable', more than 31 million were owned by approximately 7,000 landlords, whilst some 80,000 peasant families farmed less than 1.6 million hectares: 8 per cent of Bolivia's landlords controlled 95 per cent of the country's cultivable land, leaving about 5 per cent available for the remaining 92 per cent to cultivate. Large estates were growing cash crops, on average, on 1.5 per cent of cultivable land resources, whereas the 60 per cent of landowners with access to less than 5 hectares were cultivating approximately 54 per cent. The following figures indicate the virtual monopolization by the large estate system:

42

Methods of land cultivation expressed in percentages

Semifeudal cultivation	90.54
Properties worked by their owners	1.50
Properties worked with the aid of wage earners	2.44
Rented properties	2.66
Properties of Indian communities	2.86

Source: *National Statistics and Census Office*, 1950.

According to the census, although 72 per cent of all economically active people registered in 1950 were engaged in agricultural work, agricultural commodities only accounted for 33 per cent of the gross national product. The United Nations' Technical Assistance Mission Report (the Keenleyside Report) of the same year stated that:'the static nature of Bolivian agriculture has tended to impede - if not arrest - the ordinary course of economic development'. A year later, ECLA reported that the *latifundio* system was directly responsible for 'the stagnation of agriculture and its retarding effect on Bolivian economic development'. Such damning conclusions were to be acknowledged and reinforced in the preamble to the Agrarian Reform Decree (1953):

As a result of the unequal tenure of land and the defective system of exploitation characterizing it, Bolivia has a limited agricultural production, not even adequate to satisfy its internal needs, instead requiring the state to part company with 35 per cent of its financial assets, which could be invested in other urgent necessities.

In the absence of reliable data it is impossible to establish whether or not *haciendas* had achieved even the slightest increase in productivity over the years before agrarian reform. Whilst some writers glibly allude to 'the growing urban population', census figures for La Paz Department reflect the opposite trend: 1854, 471,200; 1882, 312,700 and 1900, 446,500. This implies that an increased rate of production had been unnecessary since, however unreliable the above statistics, there is no evidence for the rapid growth of an urban proletariat in the late nineteenth century. Certainly, after 1900 the population growth rate did rise but, again, there was no urgent need for the agricultural sector to increase output. Bolivia, which had been almost entirely self-sufficient in most primary foodstuffs and in many agricultural raw materials, became ever-more dependent on imported commodities until the only ones in which the country remained self-sufficient were potatoes and *quinua* (grown on the Altiplano) together with yucca and bananas (produced on the lowland plains and in the Yungas). Although agriculture was employing over 70 per cent of the total working population in 1950, 60 per cent of Bolivia's food requirements were being imported, paid for by profits from the republic's mineral industry.

Clearly, the responsibility for Bolivia's disastrous state of agriculture on the eve of the National Revolution was not entirely the estate owners'. They had been given little encouragement by successive governments to increase productivity, to experiment with new techniques or to introduce mechanization. On the contrary, in some instances they bought new properties in order to forestall competition and thus monopolize land resources by withdrawing land from cultivation: it was in their interests to restrict output i.e. to keep markets in short supply in an attempt to stimulate price increases. In this way they were being instrumental in controlling the actual level and quality of consumption within urban areas. With few exceptions, landlords were not concerned with investing capital, which could have served to reorganize the farm sector into large, mechanized labour-efficient, highly productive units: most believed that this would lead directly to a need for better educated workers and in the longer run to the evolution of a labour class more aware of their rights, thereby threatening stability and the landlords' powerful position. Under the prevailing circumstances there could never have been a flow of private investment funds which might have guaranteed agricultural development and increases in the production of agricultural commodities. With minimal investment agriculture continued to stagnate and was incapable of adequately supplying the urban population.

Exclusion of the rural masses from the country's monetary economy was equally inimical to economic development. Under the *hacienda* system peasant families survived at bare subsistence level and purchased very few, if any, goods from outside sources. Thus, an extension of the domestic market required to support and sustain industrial development did not materialize and economic progress was obstructed. A market of sufficient magnitude to promote large-scale industrialization could never have emerged in response to the limited consumption demands of estate owners alone.

Such an analysis appears static in many respects; it would be unjustifiable to argue otherwise. 'Until 1952..one could characterize the agricultural or rural sector in Bolivia as static, for no new innovations in work relations between landlord and peasant had taken place, and there had been no general acceptance of new agricultural techniques' (Clark, 1970, p.8). It would be totally misleading to apply the term 'capitalism' to the *latifundio/hacienda* system since one of its most distinctive features was a lack of capital investment. As García (1970, p.305) observed: 'Within this socioeconomic framework land was more a factor of power and social rank than a factor of production. It was not a marketable commodity, because both the old and new landowning classes lacked any sense of what a modern enterprise should be'.

4 Bolivia's Agrarian Reform Law of 1953

The National Revolution (1952)

Bolivia's National Revolution, which began on 9 April 1952, has been described as 'the most profound movement of social change in America since the beginning of the Mexican Revolution of 1910' (Alexander, 1958). Within only three days civilians and miners succeeded in bringing about the complete defeat of the military forces, presenting the victorious National Revolutionary Movement with the perfect opportunity for unleashing a radical reform programme, 'aimed at drastically restructuring existing institutions and power bases within Bolivia' (Clark, 1970, p.2). As Klein (1992, p.231) remarks, 'the new MNR leaders found themselves in total political control of the nation at a time when the elite was economically weak and incapable of opposing fundamental social and economic reforms'. Amongst the placards greeting the party leader, Paz Estenssoro, on his return from exile in Buenos Aires, were ones proclaiming: 'Agrarian Reform', 'Welcome Father of the Poor' and 'Nationalization of the Mines' (Dunkerley, 1984, p.41). Dramatic changes were set in motion almost immediately.

Whereas previously the inability to satisfy even the most modest income and literacy qualifications had precluded all but seven per cent of the adult population from participating in national politics, universal suffrage became a reality for the first time in the republic's history: the electorate increased by 1,200 per cent overnight. Shortly afterwards the army was 'so reduced in power and number that many persons believed for a time that it had ceased to exist' (Klein, 1992, p.232). At the end of April the miners, with the full support of the MNR, established a national labour union, the Bolivian Workers' Central (COB). Six months later the three leading tin mining companies of Patiño, Hochschild and Aramayo were expropriated and entrusted to a public corporation, the Mineral

45

Corporation of Bolivia (COMIBOL). The following year on 2 August - Día del Indio (Day of the Indian) - Paz Estenssoro and Nuflo Chávez Ortiz, the Minister of Peasant Affairs, together signed the Agrarian Reform Decree at Ucureña, in the company of at least 50,000 *campesinos* - both *comunarios* and *colonos*. Generally accepted as the most radical measure of the MNR's reform programme, the Agrarian Reform Law was described by the president as 'the most momentous act in the country's entire independence'. The last of the MNR's major reform acts was promulgated two years later. The Education Reform Decree declared that public education should be 'universal, free and compulsory'. The education code stipulated that the basic objective of rural education was the integration of the Indian population into national life: attainment of this ideal was to be facilitated by means of a national literacy programme and the training of skilled labour 'to increase the productivity of traditional crafts'.

Unlike the miners, the demoralized *campesinos* did not participate in the uprising of April 1952. However *comunarios* and *colonos* alike were encouraged by the content of the first speech of Hernán Siles Zuazo, acting president until Paz Estenssoro's return from exile. 'We are going to incorporate the *campesinos* into the Bolivian economy and national life so that they are no longer the objects of derision' (Dunkerley, 1984, p.41). The enacting of the Electoral Law, granting universal suffrage, gave them grounds for believing that a liberating agrarian reform programme would follow in due course. Moreover, in his first month of office, the president established a new Ministry of Peasant Affairs and Agriculture (MACA), to deal with rural matters.

Despite *campesino* expectations, in the early stages of the National Revolution MNR opinions on the future of the countryside and 'the Indian question' were almost irreconcilable. Right-wing landowners were prepared to consider the possibility of giving *colonos* titles to rented plots of land, but were totally opposed to the wholesale abolition of the *latifundio*. Left-wing party members, working in close association with the COB, were more concerned about resolving the mining situation. Dunkerley (1984, p. 66) quotes remarks made by politicians in June 1952, which contributed to the peasants' conjectures that they would only be able to achieve their goals through violent, disruptive action. ' It is not possible to proceed to the redistribution of land because this would mean establishing '*minifundia*' (very small landholdings) 'and be prejudicial to production' (Chávez Ortiz). 'I want you to lend your most decided cooperation to the government. How? By maintaining perfect order in the altiplano, producing more, working harder ... we, for our part, will do our best to improve your conditions in every sense' (Paz Estenssoro).

Klein has likened the *campesino* activities, causing 'tremendous violence and destruction' from late 1952 until early 1953 to 'the Great Fear' during the French Revolution. Violence and unrest leading to the assassination of landlords, the seizure of *hacienda* lands, the destruction of buildings, roads etc. was, again, centred on Cochabamba and the Lake Titicaca region. 'Around the shores of Lake Titicaca and in the Cochabamba valley *hacienda* buildings were attacked weekly and although these petty revolts dissolved through lack of direction or sustained support, they often included large numbers of rural labourers' (Dunkerley, 1984, p.67). Within several months 1,200 peasant *sindicatos* (syndicates), with 200,000 members, had become active in the Cochabamba area alone. Lindqvist (1974, p.116) refers to a rare example in spring 1953 of the police chief at Achacachi (the provincial capital of Omasuyos) ordering his men to fire on 'peasants who refused to bow to the landowner'; in most cases landlords under serious threat retreated to their urban residences.

The Minister of Peasant Affairs had personally initiated the campaign, encouraging *campesinos* in different parts of the country to group together in peasant syndicates. Although he tried to assure urban residents that there was 'no agitation' and that field tasks were proceeding normally, clearly this was not the case. Undoubtedly, it was the fear of further violence and destruction, together with the threat of losing control, that persuaded - if not forced - the MNR early in 1953 to begin planning in earnest for the implementation of agrarian reform.

Motivating forces

A number of researchers have, over the years, questioned the MNR's commitment to radical reform, especially in connection with the abolition of the *latifundio*. Some have been convinced that rural peasant violence during late 1952 and early 1953 forced both the pace and scope of change. Thus Huizer (1973, p.56) maintains that: 'The new government saw that only an overall land reform eliminating the *hacienda* system could channel and control the rising tide of peasant unrest'. Likewise Klein (1992, p.234) alleges that: 'However reluctant the new regime may have been to attack the *hacienda* problem seriously, the massive mobilization of the peasants, now the majority of the electorate, and the systematic destruction of the land tenure system forced the regime to act'. It has been suggested that, since both Paz Estenssoro and Chávez Ortiz were members of landowning families, it was in their interests to press for 'the right of property'. Hence agrarian reform before 1952 had only been 'discussed in the most general of terms, usually with reference to a vague 'incorporation' of the *campesinado* into a pre-existing society and nation' (Dunkerley, 1984, p.65).

What were the principal motives behind the introduction of agrarian reform? Were the MNR leaders, especially Dr. Paz Estenssoro, a respected economist, primarily concerned about social justice or was economic efficiency the overriding objective? Osborne working in the British Embassy in the year preceding the 1952 insurrection, had no doubts: 'The MNR party came to office pledged to restore the land to those who work it ... It is unnecessary to doubt that this reform was motivated by genuine idealism, by the desire to right the wrongs of the past and to integrate the Indian peasantry more closely into the fabric of the state'. He went on to suggest that 'it was also a particularly astute political move', in that it secured a majority support for the party (1964, p. 74). Carter Goodrich (1971, p. 21), a US economist, who witnessed the uprising of April 1952, was convinced that 'the main motives for reform were moral and political, to redress injustice and to redistribute power'.

The president's preamble to the Agrarian Reform Law began with an acknowledgement of the agricultural achievements of the Incas and condemnation of the ways in which the Spanish Conquest and colonialism had ruthlessly and violently dislocated the agrarian economy and enslaved the indigenous people. Two of the preamble paragraphs illustrate the perceived linkages between social justice and economic efficiency.

Considering: That in spite of the material and spiritual protection of the Law of the Indies, the Indian race, by the imposition of a semi-feudal system, **with the *repartimientos* and *encomiendas*, was unjustly stripped of property and subjected to personal and gratuitous serfdom**, establishing for the first time the problem of the Indian and the land, not as a racial or an academic problem, but essentially as a social and economic one.

Considering: That the National Revolution, in its agrarian programme, is proposing essentially to raise the country's actual production level, to transform the feudal system of tenancy and exploitation of the land, putting in plan a just redistribution amongst those who work it, and to incorporate in national life the Indian population, regaining possession of their economic hierarchy and human condition.

Interviewed by the author on two occasions in 1971, Paz Estenssoro maintained that it was 'impossible to separate agrarian reform from politics'. He made it abundantly clear that, since the formation of the MNR, 'the liberation of the *campesinos*' had been at the forefront of the party's philosophy, though he added that exactly how far one should put personal freedom before national production posed 'an academic question'. Throughout the deliber. 'ions on agrarian reform he had remained convinced of the necessity to abolish 'the feudal *latifundios*' for both moral and economic reasons, but simultaneously determined to preserve existing modernized, highly productive agricultural enterprises.

Paz Estenssoro was adamant that, whilst he had expected the implementation of agrarian reform would bring about a radical restructuring of agriculture, he had never envisaged that the 1953 legislation would be an end in itself; he had viewed it rather as the first stage of an ongoing process of agricultural adaptation and modernization. From the outset the MNR had also anticipated a fall in national agricultural output during the period of adjustment but expected productivity to increase by the end of the decade.

The Agrarian Reform Commission

The Agrarian Reform Commission was appointed on 20 January 1953, though its complete membership was not confirmed until late March. Chaired by Siles Zuazo, the vice president, the Commission, which began formal discussions at the beginning of April, was given the daunting task of drafting proposals for a thorough-going programme of agrarian reform within a period of three months. On 28 July the Commission presented its very lengthy report and five days later the Agrarian Reform Law (Decree Law No. 3464) was signed at Ucureña by the president and all the members of the cabinet, in the presence of a vast gathering of *campesinos*. Significantly, the first national peasants' conference had been held in La Paz in June and on 15 July the Bolivian National Confederation of *Campesinos* (CNTCB) had been founded to promote the creation of federations and syndicates in provinces where they had not previously been formed.

The Commission's members represented political, economic and agricultural interests. Whether or not in order to guarantee control of *campesino* insurgents (as suggested by Dunkerley), an early decision was taken by the MNR to invite delegates from the two main left-wing parties i.e. the Revolutionary Workers' Party (POR) and the Party of the Revolutionary Left (PIR), to participate in the discussions. As a consequence, Arturo Urquidi, leader of the PIR and an agricultural expert, came to dominate the work of the Commission: 'The final draft of the decree bore the stamp of the PIR's objective of developing capitalism in Bolivian agriculture on the basis of medium-sized owner-worked or cooperative holdings as an essential stage before widespread collectivisation could be introduced' (Dunkerley, 1984, p.72).

When interviewed in 1996, Roberto Jordan Pando, a sub-secretary in the Ministry of Peasant Affairs in 1953 and later to become Minister of Peasant Affairs and Agriculture, recalled the period of consultation on agrarian reform as one of excitement, enthusiasm and euphoria: the Mexican *ejido* and the recently created Chinese cooperative were avidly studied in the quest for a 'national', 'liberal', 'democratic' 'campesinization' model of reform. The Commission eagerly sought the participation and professional advice of academic experts and agricultural economists familiar with well established agrarian reform programmes and their administration. Flores (see p.2), appointed as an agricultural adviser by the UN, remained convinced that in order to be effective reform had to fulfil three conditions:

49

1 It has to take productive land and its income, above a ceiling which is exempt from the reform and is determined by political considerations disguised as economics about the optimum size. Productive land must be taken *without immediate compensation*. Otherwise it is not a redistributive measure. To claim that landlords should be fully compensated is as absurd as to expect that taxpayers of advanced countries should receive cash compensation or bonds in an amount equal to their taxes.

2 It must take place rapidly and massively: say, within one or two decades. Otherwise it will not generate the momentum for take-off. Instead, it will depress even further the performance of the agricultural economy and set in a process of disinvestment because of the spread of uncertainties.

3 It must be accompanied by vigorous development policies within agriculture and outside it. In the agricultural sector a new, flexible and efficient pattern of resource allocation and use must be created. Simultaneously, there has to be a transfer to industry and trade of capital originally tied up in land (Flores,1970, p.152).

Whilst there was by April 1953 unanimous agreement amongst Commission members about the need to abolish the *latifundio* system and liberate *campesinos* from their condition as serfs, compromises had to be reached on more contentious issues. Not surprisingly, landholding Commission members could only accept with considerable reluctance Flores' views on compensation payment to *hacienda* owners whose lands were to be expropriated. Clark (1970, p.90) has stressed the crucial role played by the president in maintaining harmony: 'the personality and ideas of Paz Estenssoro were to form the primary criteria, after close discussion with the committee'. He expressed similar views to Paz Estenssoro about the inextricable links between 'Latin America's most dynamic social and economic revolution since the Mexican Revolution' (Klein, 1992, p.226) and the radical model of agrarian reform contrived by the Commissioners:

An informant has stated that perhaps the most important factor in this harmony, despite varying opinions and backgrounds of members, was that the revolution itself was already a reality; it was backed by the new president with undeniable national support from the mining and peasant sectors. Even the most conservative members of the committee could not oppose this (Clark, 1970, p.90).

Few agrarian reform laws have been drafted in such auspicious circumstances. The government's rationale for the implementation of a comprehensive programme of agrarian reform was set out in the preamble to the Agrarian Reform

Law: it bore the clear imprint of the president. The *latifundio* was described as an 'inhumane' and 'degrading' institution - the principal cause of past 'bloody insurrections' to recover usurped lands. Since the last decade of the nineteenth century, *latifundistas* (the owners of *latifundios*) had operated in close association with mining barons. Whilst mining had diverted financial investment from national agricultural production, the *latifundio* system had demonstrated its 'incapacity to evolve in accordance with the historic necessities of the country'...it had instead become 'an obstacle to the country's progress'. It was stated categorically that lands usurped from Indian communities should be expropriated and become the property of 'those who work them'. Previously uncultivated lands seized illegally by landlords should revert to state ownership for colonization purposes, to satisfy needs arising from immigration, and for other public uses: property should 'serve a social function'. On the other hand, under the new landholding structure, the rights of peasant farmers and medium-sized property owners should be respected.

Because of the 'unjust' and 'irrational' distribution of agricultural property, and the *campesinos'* lack of protection by previous governments, making it impossible for them to subsist in their communities of birth, a process of ever-accelerating migration to the cities, to mining centres and neighbouring countries had become established, causing incalculable damage both in terms of the demographic interests of the nation and agricultural output. Usurpation of Indian property and imposed servitude were largely responsible for an adult illiteracy level of 80 per cent, for a total lack of technical education for peasant farmers and the widespread 'contempt for artistic traditions', the values of national folklore and the ethnic qualities of the native worker. Such a state of servitude, and consequent 'backwardness and ignorance' were, it was stated, responsible for sub-standard housing conditions and unacceptable levels of nutrition and health. It was anticipated that the process of agrarian reform would be accompanied by a whole series of initiatives to improve the *campesinos'* living standards and enable them to participate in *'la vida nacional'* (national life). In essence, the basic objective of agrarian reform was to 'transform the feudal land tenure system by promoting a more equitable distribution of land, raising production, and integrating the rural population into the national economy and society' (Clark, 1970, p.32).

The six fundamental objectives of the Agrarian Reform Law

The preamble concluded with a statement of the six 'fundamental objectives' of the Agrarian Reform Law, as agreed by all members of the Commission and signed by the president:

1 To allot cultivable land to the *campesinos* who do not have any, or have very little, providing they work it; expropriating for this purpose the lands which *latifundistas* hold in excess, or from which they enjoy an absolute rent, not earned by their personal work in the field.

2 To restore to indigenous communities the lands that were usurped from them and to cooperate in the modernization of their agriculture; respecting and making use, where possible, of their collective traditions.

3 To free *campesino* workers from their conditions as serfs, from gratuitous personal services and obligations.

4 To stimulate greater productivity and commercialization of the agricultural industry facilitating the investment of new capital, respecting small and medium-sized farms, encouraging agrarian cooperation, lending technical aid and opening up possibilities for credit.

5 To conserve the country's natural resources, adopting indispensable technical and scientific means.

6 To promote the internal migration of the rural population, now excessively concentrated in the inter-Andean zone, with the aim of obtaining a rational human distribution, of strengthening national unity and of integrating the eastern area with the western area of Bolivia.

The main provisions of the Agrarian Reform Law

Although *Decreto Ley No. 3464*, containing in all 177 articles in 16 sections, was detailed, complex and, in parts, somewhat obscure and ambiguous, its main provisions were clearly and categorically set out.

Types of agricultural holdings

The document stated at the outset that, whilst 'the soil, the subsoil and waters of the republic' belonged to the Bolivian nation, the state recognized and guaranteed 'the private agricultural property when it' was 'carrying out a useful function for the nation's population'. From that time onwards only six types of agricultural landholdings would be legally recognized:

1 *El solar campesino* (the *campesino* ground plot/residential plot), considered 'insufficient for the subsistence needs of a family'.

2 *La propiedad pequeña* (smallholding), worked on a subsistence basis by the *campesino* and his family.

3 *La propiedad mediana* (medium-sized holding), 'which without the characteristics of the capitalized agricultural enterprise is farmed with the aid of wage-earners or technical equipment' in such a way that 'the bulk of its produce' is 'destined for the market'.

4 *La propiedad de comunidad indígena* (indigenous community property), recognized in law as indigenous community land and farmed for the benefit of the entire community.

5 *La propiedad agraria cooperativa* (cooperative property), worked jointly as a result of: (a) farmers grouping together in order to obtain farm land and exploit it; (b) owners of smallholdings and medium-sized properties working together; (c) *campesinos* gaining legal rights to former *latifundio* land and deciding to farm it cooperatively or (d) cooperative society members owning the land for other reasons.

6 *La empresa agrícola* (agricultural enterprise), 'characterized by large-scale capital investment, a paid work force and the use of modern, technical methods, except in difficult topographical regions'. As Flores and Goodrich were to observe, the creation of the *empresa agrícola* was 'quite unique in land reform legislation' (Goodrich, 1971, p.21).

On the other hand, Article 12 and subsequent clauses declared very forcibly that '**the state no longer legally recognizes the existence of the *latifundio*** characterized by its great territorial extent, its lack of, or inefficient, exploitation, antiquated farming methods' creating 'a regime of feudal oppression, resulting in backward farming and a low level of life and culture for the *campesino* population'. According to Article 30, not only was the *latifundio* abolished but also declared illegal were other forms of extensive landholdings owned individually or corporately.

Articles 13 to 18 listed the maximum sizes permissible for smallholdings, medium-sized properties and *empresas* 'in accordance with geographical zones'. Thus, the maximum extent of a smallholding in the most fertile and climatically favoured part of the Zona de Valles was given as only three hectares, whereas in the arid sub-zone of the Chaco it was as much as 80 hectares; likewise, the maximum permitted size of the medium-sized property ranged from 24 to 600 hectares. Whilst an agricultural enterprise in the most productive zones could not exceed 80 hectares, such a property could legally reach a size of 2,000 hectares in the tropical and sub-tropical zones of the Oriente. An *empresa ganadera* (capitalized cattle ranch) in one of the same zones could cover up to 50,000 hectares of land if the landowner grazed at least 10,000 cattle; otherwise, he was to be allocated 5 hectares per cow.

The difficulties, complexities and potential abuses associated with determining the maximum extent of different types of property in ill-defined geographical zones are indicated by the data provided for the Altiplano. In this case the confusion was increased by inconsistencies in the zoning between property types.

Article 15: Smallholding

North sub-zone, shores of Lake Titicaca	10 has
North sub-zone, with the influence of Lake Titicaca	10 has
Central sub-zone, with the influence of Lake Poopó	15 has
South sub-zone	35 has

Article 16: Medium-sized property

North sub-zone, with the influence of Lake Titicaca	80 has
North sub-zone, without the lake's influence	150 has
Central sub-zone	250 has
South sub-zone and semi-desert	350 has

Article 17: Agricultural enterprise

Zone influenced by Lake Titicaca	400 has
Andean zone and Altiplano	800 has

Land expropriation and redistribution

A clear distinction was drawn between agricultural lands whose ownership was declared to be inviolable and those designated for expropriation and redistribution. Small properties, provided they did not exceed the maximum size stipulated for their area, naturally fell into the first category. Whilst the ownership of some medium-sized landholdings presented no problems, on other properties *campesinos* were entitled to the *sayañas* (house plots) they had previously cultivated for their own benefit; in some parts of the country the seizure and possession of such plots was already a *fait accompli*. Where such plots were left 'vacant' as, for example, when *campesinos* were legally allocated other *parcelas* (patches of land), they could be 'consolidated' by the owner of the medium-sized property, providing the *campesinos* in question were compensated. A *campesino* whose house was situated on the land of either a medium-sized property or an agricultural enterprise had the legal right to continue living in it until such time as the property owner constructed, at his own expense, a house of the same value and built of the same materials, on the private land of the *campesino*.

Whilst the law protected the *hacienda* house, the typical *latifundio* was subject to expropriation in its entirety. On the other hand, *hacendados* who were able to prove that they had farmed 'efficiently' (by investing capital, employing waged labour and adopting modern machinery and techniques), were entitled to have their holdings reduced to the dimensions of medium-sized properties, 'with all the rights and duties applicable to medium-sized landholdings'.

For all *campesinos* two articles in the bill's section on ownership rights were highly significant. Article 42 restored, after due process of law, all lands usurped by unscrupulous landlords from *comunidades indigenas* since 1 January 1900. Article 60 fulfilled the third of the six 'fundamental objectives' in that it dealt a final death blow to *pongueaje*, declaring all forms of obligatory personal services and contributions illegal. If political, military, municipal or ecclesiastical authorities exacted such contributions after the promulgation of the Agrarian Reform Law, they would be committing a flagrant crime. Later clauses referred to the abolition of the *colonato* system and unwaged labour arrangements and the cancellation of all debts to *hacendados*. The *campesino* worker was to be incorporated into 'the Nation's legal-social system, with all the rights recognized by the law'.

The question of entitlement to expropriated agricultural land had been debated at length by the Agrarian Reform Commissioners. According to the final version of the Decree Law, 'all Bolivians over the age of 18, regardless of their sex', who dedicated themselves to agricultural pursuits, or who wished to do so, would, after due process of law, be allocated land where available 'in accordance with the government's plans' (Article 78). In an endeavour to forestall traditional problems of 'idle land' and absentee landlords, it was clearly stipulated that the beneficiaries of *dotación* (land settlement) would only be allowed to retain land assigned to them if they displayed evidence of working it within two years.

All *ex-colonos* who had been 'subjected to a feudal regime of work and exploitation', including 'married men over 14 years of age and widows with young children', were declared to be the owners of the plots of land they actually possessed and farmed. If *haciendas* were large enough to enable all community *campesinos* to benefit from redistribution, any excess land could be allocated to less fortunate *campesinos* in other communities within a radius of 10 km. On the other hand, if land was in short supply - as was frequently the case in the densely populated communities of the northern Altiplano-*campesinos* would be entitled to petition for land grants in unspecified '*otras areas disponibles*' (other available areas). *Campesinos* who already had legitimate access to plots of land, would be allocated just enough *hacienda* land to create parity amongst community members. To foster a spirit of cooperation, at least 10 per cent of redistributed lands would be designated for collective usage, in addition to an area set aside for school buildings and recreational purposes.

According to Article 121, a policy of 'regrouping' arable fields would be applied in areas where the *minifundio* predominated i.e. areas in which 'the majority of properties' were 'of insufficient size to ensure the subsistence of their owners, whose main occupation' was agriculture. Cooperatives of *minifundistas* would be given priority, with access to the main communication lines, 'in their nearest colonization zones'.

Emphasis was laid on the fact that the phrase 'all Bolivians over the age of 18' did not apply solely to rural dwellers: on the contrary, urban populations were given every encouragement to relocate in the countryside. Surplus expropriated lands in the Yungas were to be made available in the form of small properties for redistribution 'to invalids and relatives of those killed in the National Revolution, to all middle class Bolivians, employees and professionals, railway workers and builders, equally to all factory workers and miners' who wished to settle there.

Clearly, *dotación* and the legal granting of titles took very different forms. Carter (1971b, p.246) has identified five categories: 'affectation, restitution, consolidation, inaffectability, and outright grant'. Affectation (to become the most widespread and least straightforward form) involved the expropriation and redistribution of *latifundio* lands. Restitution referred to the restoration of lands usurped from Indian communities between 1900 and 1953. Consolidation was the term applied to the legal titling of lands 'in cases of small and medium-sized properties where no *colonato* system had ever developed' but the occupants were desirous of obtaining legal titles. The state of inaffectability was 'declared primarily in cases of capitalized and mechanized agricultural enterprises' where the property owners wanted 'the legal security of a clear title as well as a defence from encroachment by squatters'. Outright land grants from land held in *el dominio público* (the public domain) would be restricted mainly to the sparsely populated regions of the Oriente. In all five cases, the same conditions applied; beneficiaries had to work the land within two years of receiving titles or forfeit all legal rights; properties could not be sold and, if an owner died without a legal heir, his property would be reclaimed by the state.

The numerous and complicated procedures involved in the actual process of *dotación* were omitted from the Agrarian Reform Law; instead they were described in detail in a lengthy supreme decree, *Servicio Nacional de Reforma Agraria* (National Agrarian Reform Service: SNRA), published at the end of August 1953. Further decrees, such as one on the duties and payment of land surveyors, were drafted as and when problems arose and points required clarifying.

56

According to the Agrarian Reform Law, before formal measures could be taken to set in motion the adjudication process, *campesinos* or the landowner, 'separately or together', had to present a *demanda* (a land claim petition) to the National Agrarian Reform Service, the institution responsible to the Bolivian president for administering agrarian reform. In response, the SNRA would authorize a comprehensive survey of the property to ascertain precisely which lands were available for expropriation and the number of claimants concerned. Details about land to be assigned for collective usage were to be agreed with the community or *campesino* syndicate. Until national agricultural survey plans were published, any compensation arrangements would be based on a five-grade land classification system. Land quality ranged from 'first class' (with fertile soils, a favourable climate and plentiful water supply, enabling two crops to be harvested annually), to 'fifth class' i.e. the poorest sandy or rock-strewn terrain unsuitable for any type of farming activity.

Not surprisingly, compensation had remained a thorny issue throughout the Commission's deliberations. The views of Chávez Ortiz and other Commissioners with landowning interests were markedly different from those of the PIR and POR representatives, adamant that, since many *haciendas* had originally been seized from their rightful owners, *ex-hacendados* could hardly expect any reimbursement for their loss of land. The more moderate opinions of Flores et al. (that a modest amount of compensation should be paid over a lengthy span of time) eventually prevailed.

Ex-landowners were to be paid compensation in the form of agrarian bonds bearing a simple interest at two per cent and maturing in 25 years. In the event, disputes about compensation were to prove meaningless. As Clark observed in 1970 (p.41):

No official compensation has ever been paid; nor have the agrarian bonds been issued. Notably, the ex-landlords have not tried to enforce this provision, probably because the compensation and bonds were to be based on the landlords' own declaration of the properties' values [inevitably unrealistically low] for tax purposes.

Rampant inflation had the effect of making compensation worthless, since the Agrarian Reform Law had failed to provide landowners with any form of protection against it. Likewise the *campesino* beneficiaries 'were never obliged to assume part of the costs for compensating landlords' (as required under Article 160), 'since the Government never acted on this matter' (Clark).

Colonization

Although colonization as such was not specifically mentioned in the last of the six 'fundamental objectives', it was clearly implied in the reference to 'internal migration'. Despite Flores' strongly held views about the 'bad economics' and 'inadvisability' of 'opening public domain lands before industrial development gets under way' (1970, p.149), colonization was presented as an inevitable and integral component of Bolivia's agrarian policy. The republic's colonization programme, although it proceeded piecemeal during the early stages, did in fact pave the way for the planned re-settlement schemes introduced throughout the 1960s and 1970s in Brazil, Colombia, Ecuador, Paraguay, Peru and Venezuela. Morris (1987, p. 44) reflects the opinions originally expressed by Bolivia's Agrarian Reform Commissioners: 'As an alternative to agricultural development in the *altiplano*, colonization here is more justifiable than in other countries, simply because there is no more good land available in the highlands'.

Whilst writing colonization into the Agrarian Reform Law, MNR members made it abundantly clear from the outset that the party's eagerness to 'open up' the Oriente, by establishing settlement colonies, was motivated by more than agricultural considerations. 'Strengthening national unity' and 'integrating the eastern area with the western area of Bolivia' were vital concerns in the government's strategic planning, particularly as a number of politicians had fought in the Chaco War, which had ended in such ignominious defeat and with the devastating losses of both territory in the formerly neglected south-east and the country's highly prized outlet to the Atlantic, by way of the River Paraguay. Dickenson et al. (1996, p.143) maintain that: 'This military-strategic motivation, based on the principle that occupation of territory will validate political frontiers in sparsely populated territory, seems to underlie a good deal of thinking about colonization in Latin America'.

It was anticipated that a rational distribution of population would be the natural outcome of a carefully designed and implemented colonization programme. Not only would this relieve population pressure on the Altiplano, resulting in *minifundismo*, landlessness and, ultimately, destitution, but it would also provide a desperately needed agricultural labour force in the sparsely peopled vast expanses of the Oriente. Carter (1971b, p.255) observed that at the time of the National Revolution, 'the value of labor was, proportionately, so high that a Bolivian economist' (Sanjinés) could write: 'Santa Cruz is one of the few places in the world where the work of a farm laborer is much more valuable - within a short time - than a hectare of land'. The departments of Santa Cruz, Beni and Pando, together accounting for two thirds of the republic's land surface, but with only a fifth of its population, contained extensive lowland plains ideally suited for the production of tropical and sub-tropical crops, both for commercial purposes and peasant consumption.

Writing originally in 1954, Osborne referred to the Oriente as being 'underpopulated, underdeveloped, very deficient in means of communication, especially in the northern and larger section ... in danger of becoming isolated from the rest of the country in consequence'. Yet this was the area containing Bolivia's 'extensive sub-Andine oil belt running northwards from the Argentine frontier .. in the direction of Santa Cruz and thence turning north-westwards along the Piedmont of the Cordillera Real to the frontier with Peru'. In November 1952 the national oil corporation (YPFB) had strongly recommended the government to expand its exploration operations: as a result of doing so, whereas 'up to 1954 Bolivia imported petrol and petroleum products to an annual value of about $8 million ... by the autumn of 1955 the Government was able to announce that the output was adequate for domestic needs and exports exceeded $6 million a year' (Osborne, 1964, pp.156-158). Clearly, as Thiesenhusen (1995, p.65) summed up the situation, 'in its quest for growth, the government realized the potential of the lowlands for earning foreign exchange'. In the MNR's determination to proceed with colonization projects, the urgent need for improving inter-regional transport systems and widening market facilities was fully appreciated. In 1954 the completion of a highway (referred to by Osborne as 'the most important road ever built in South America') directly linked Cochabamba with Santa Cruz, at that time 'a backwater cattle-producing town' (Swaney and Strauss, 1992, p.313). The following year work was also completed on the railtrack between Santa Cruz and Corumbá, on the Brazilian frontier.

The Agrarian Reform Law made no reference to different types of colonization (directed, semi-directed or spontaneous), nor did it stipulate how many upland families should be re-settled. The details were more sharply defined in 1960, when the first directed colonization projects were implemented in three areas: the Alto Beni (north-east of La Paz); the twisting valleys of the Chapare (between Cochabamba and Santa Cruz) and the Santa Cruz region. (In the case of the Chapare, nowadays internationally known for its coca production, spontaneous colonization by Quechuan Indians from the valleys of the Cochabamba area had occurred intermittently from the turn of the century). Articles 114 and 116 stipulated that 'first class zones' were to be selected as appropriate areas for settlement: whilst two thirds of the land in question would be allocated for small and medium-sized properties, the remaining third would be set aside by the SNRA for the creation of agricultural enterprises. Three 'first class zones' were identified: (1) a belt of uncultivated land, or land restored to the public domain, up to 25 km in width on each side of main communication lines, including navigable waterways; (2) within the tropical and sub-tropical *llanos* (i.e. between the *selva* (forest) and arid lands of the Gran Chaco), land up to a distance of 5 km around nucleated settlements containing at least 1,000 residents and (3) such other lands as might be designated by the government in future years.

In these zones preference would be given to landless *campesinos*, unemployed workers, Bolivians returning from abroad to live in the country, veterans of the Chaco War and relatives of those killed in the National Revolution. The offer of land grants was also extended to foreigners wishing to settle in colonization areas. Thus, by the 1960s, communities of Okinawans, Japanese and Mennonites from Paraguay were to become firmly established in the department of Santa Cruz.

Additionally, Article 91 stated that all *campesinos* from the Altiplano and Valles zones were entitled, 'independently' of any other landholdings, to legally own 50 hectares of cultivable land 'in the region of the Oriente', provided they satisfied the normal requirements, such as tilling the land within the first two years of occupation. According to Paz Estenssoro (1971), the only incentive *campesinos* had needed to persuade them to move eastwards had been 'the promise of available land' - a viewpoint challenged later.

Conservation of natural resources

The Agrarian Reform Law's approach to the conservation of the country's natural resources was, for the most part, enlightened for the early 1950s. Unfortunately, no realistic methods of enforcing proposed policies were suggested. Likewise, 'the indispensable technical and scientific means', referred to in the fifth of the 'fundamental objectives', were not forthcoming.

Significantly, state ownership of 'the soil, subsoil and waters' within the republic's territory was affirmed in the first clauses of the Agrarian Reform Law. Articles 3 and 4 emphasized that all *tierras baldias* (uncultivated lands), *tierras vacantes* (vacant or empty lands), lands belonging to state organizations and authorities constituted state property, together with lakes, lagoons and rivers.

Chonchol's fifth of 'eight fundamental conditions of agrarian reform in Latin America' (see p.12) stipulated that 'agrarian reform must affect both the land and water ... Even though the constitutions generally establish that water is a national wealth for public use, in practice most agriculturalists act as though water were a private asset over which those who own land can exercise the power of use and abuse' (1970, p.166). Section eight of the Agrarian Reform Law stated categorically that all populations had the right to *agua potable* (drinkable water) - sadly, still not a reality in a large number of rural communities in Bolivia. Additionally, 'as a general rule, the water entering a property' was 'to be used as required by the owner, without anybody impeding its use for agriculture'. Water resources could not be sold commercially but it was incumbent on property owners to make surplus water available to water-scarce areas or individual landowners.

Article 146 was specifically concerned with 'the conservation of forest resources and animals'. Plant resources, such as quinine-yielding cinchona, precious timber resources, animals and birds (in danger of extinction by reason of their valued skins and exotic plumages) would be placed under national protection: the state would 'organize and rationalize their exploitation'. One brief section of the law dealt exclusively with the management of rubber and chestnut trees in the Oriente. All such 'plantations' would revert to the public domain. Rubber tappers, who had previously roamed the forests to gather resin at will, were to have their access restricted to two small designated areas.

As a means of averting land degradation, the 'destruction' of woods and pastures was declared illegal on all lands with an incline 'exceeding 15 per cent'. A somewhat ambiguous clause authorized farmers in the Yungas and densely populated areas to extend the land available for cultivation purposes by terracing. An undertaking was given that the state would in future supervise animal husbandry i.e. the rearing of cows, horses, mules, sheep, pigs, goats and camelids: records would be kept of total numbers, rearing methods, animal health and marketing - a virtually impossible task. Laws concerned with the conservation of soils, on national parks, forestry and fish would be published in due course. Surprisingly, for a country so heavily dependent on the export of minerals, there was no reference to the ownership of, or the need to conserve, mineral resources.

The *pueblos indigenas* of the Oriente (including the Ayoreo, Chiquitano, Chiriguano, Garavo, Chimane and Mojo peoples) were dismissed rather patronizingly in one sentence (Article 129). 'The forest groups of the tropical *llanos*, who are to be found in a savage state and have a primitive form of organization, remain under the protection of the State'. As Wearne (1996, p.119) comments:

> Many [Latin American land reform programmes] involved transmigration, settling landless indigenous people from highland regions in lowland tropical forest zones on the false assumption that such land was both empty and suitable for subsistence agriculture. In Peru, Bolivia, Guatemala and Mexico such programmes often sent one dispossessed highland indigenous group to dispossess another lowland group.

Campesino communities, syndicates and cooperatives

Preamble references to 'the collective traditions' of indigenous communities and 'agrarian cooperation', together with the fact that by mid-1953 rural syndicates, fully supported by the COB, had ousted a significant number of *hacendados* from their properties, made it inevitable that peasant organizations would feature prominently in the text of the Agrarian Reform Law. The ninth section examined in turn the roles *campesino* communities, rural syndicates and agrarian

61

cooperatives were expected to play in order 'to stimulate greater productivity and commercialization of the agricultural industry' (objective four).

La comunidad campesina (peasant community), guaranteed legal protection, was defined in broad terms as 'a group of people bound together spatially and by common interests'. Communities could be categorized according to their origin. The *comunidad de hacienda* (*hacienda* community) consisted of at least 50 families, whose members had been subjected to exploitation under the large estate system. Together they had 'constituted a unit of production with the day-to-day discipline of collective work'. The *hacienda* community would be given the opportunity 'to maintain the system of cooperative production followed on the *hacienda*' but would be working for the benefit of the community's families rather than for a landlord. A *comunidad campesina agrupada* (a grouped peasant community) was composed of families farming small and medium-sized properties, voluntarily grouped together in order to obtain legal recognition as a settlement. A *comunidad indigena* (sometimes referred to as a *comunidad originaria* i.e. original community) comprised the families of *originarios* (descendants of original settlers) and *agregados* (land-possessing later settlers), 'living in and farming an area of land legally recognized as community land by reason of titles being granted either during or after the period of Spanish colonialism'. Regardless of its origin, a *campesino* community had a number of functions it was expected to fulfil:

(a)   to represent the interests of its members in legal matters;
(b)   to promote the well-being of its population, particularly with respect to:
    1   education;
    2   the improvement of living conditions;
    3   health care and protection;
    4   the improvement of production techniques;
    5   the promotion of forms of cooperation to produce the necessary economic resources and provide the personal labour essential for undertaking development projects.

Article 59 advised the *campesinos* of *comunidades indigenas*, many with problems of land fragmentation as a result of inheritance rights, to bring about 'the rational usage of land', by negotiating with state assessors for the plots to be regrouped. Redistributed *hacienda* land designated as collective land was to be exploited by all community members: profits were to be used for the benefit of the community.

It was stated that the *campesino* community differed from the agrarian syndicate in that it did not engage in 'class struggles against outside forces' and could not be 'a component of provincial, departmental or national organizations'. At the same time it was emphasized that conserving its independence did not exclude the existence of agrarian syndicates and other cultural, economic and political organizations.

Article 132 confirmed the *de facto* situation regarding *campesino* syndicates. There was no denouncement of syndicates for assassinating landlords, or for the sort of activities described by Patch (1961, p.129): 'Land, buildings, seed stocks, animals, vehicles, and machinery were seized and divided among syndicate members'. Neither was it affirmed that the main role of the agrarian syndicate was to promote increased agricultural productivity; the rhetoric was confined to socio-political and legal issues. The syndicate was recognized as an 'instrument for defending the rights of its members and for protecting social gains'. Syndicate members were expected to play a prominent role in activating agrarian reform: as in Mexico, without their presentation of petitions for land redistribution, formal proceedings could not be initiated. As Lindqvist (1974, p.118) commented: 'The law recognized the syndicates' right - yes, even their duty - to participate in the land reform. In practice these syndicates became the most important tool for the reform's implementation'.

It soon became apparent that the *sindicato* would replace the *hacienda* as the central mechanism of social control and become the principal interface between government and the rural masses. The MNR, and Paz in particular, recognised the great potential of this and would turn to the peasantry over the following years as a dependable ally in the struggle against the militant miners (Dunkerley, 1984, p.74).

Dunkerley quotes a statement made by the newly appointed Minister of Peasant Affairs, Chávez Ortiz, in the early days of the National Revolution: 'We are going to orientate the Agrarian Reform on the basis of strengthening collective communal property and the implementation of the capitalist stage in private property, liquidating feudalism'. Not surprisingly, such controversial comments - made by the son of a large landowner - had given rise to premonitions amongst right-wing, landholding politicians of landlord denouncement meetings and summary killings (rife at that time in China) and the introduction of a policy of wholesale collectivization. Chávez Ortiz, forced to clarify the MNR's intentions, stressed that 'the new system of agrarian labour' would 'be developed on the basis of the existing indigenous communities in order to bring about cooperativisation'.

According to Paz Estenssoro (1971), working cooperatively in the fields for the benefit of both family and community and centuries-old forms of reciprocal obligations (especially *ayni*) were deeply engrained features of rural community life: they had provided a sure foundation on which to build more formalized cooperative structures. However, whilst agrarian cooperation (through price bargaining, bulk purchasing of seeds, fertilizers etc.) could stimulate agricultural productivity, he had been convinced that communal ownership of property - as was the situation in Mexico (where land reform had enabled peasant farmers to work *ejido* land individually, but land was held on a communal basis) - would not be appropriate on the Altiplano. It would have a bad psychological effect on the *campesinos*, to whom individual land titles were extremely important and highly prized: few *campesinos* having gained coveted titles to land parcels through *dotación* would wish to exchange them for membership of an impersonal landholding organization. Exactly how many Commissioners dissented from Paz Estenssoro's views on the traditional nature of cooperation is unknown. Article 133 endorsed the social value of the agrarian cooperatives: the state would promote their organization and development. Cooperatives would be founded on a number of basic principles: open membership; equality in rights and obligations; democratic control and the single vote, regardless of members' capital input; distribution of profits according to the quality and quantity of work carried out or the number of projects completed.

Agrarian reform?

The clauses of the reform law reviewed so far were concerned for the most part with changes in the landholding structure, labour relations and peasant organization. What of agrarian reform? A number of researchers have, over the years, written indiscriminately about Bolivian 'land reform' and 'agrarian reform'. Clark's Country Paper for US AID (June 1970) was entitled *Land Reform in Bolivia*: likewise, Lindqvist's chapter in *Land and Power in South America* contained a section on 'The Land Reform'. It has even been suggested that, because of the lack of a separate term in Spanish for land reform, the *Reforma Agraria* in the *Ley de Reforma Agraria* really referred to land reform. Flores, himself, added to the general confusion. Before his article *Un año de reforma agraria* (A year of agrarian reform) appeared in a Mexican journal in 1955, he had published a similar one in English with the title 'Land Reform in Bolivia'. In a 1965 article, with the title 'The Economics of Land Reform', he had remarked that: 'It is not surprising that today practically all political programs in underdeveloped nations should consider agrarian reform one of their basic objectives' (Flores, 1970, p.142).

Unquestionably, the intention of Paz Estenssoro, the MNR and the majority of Commissioners was to introduce an all-encompassing agrarian reform programme; their policy aims were clearly set out in the preamble's six fundamental objectives. In drafting the proposals for agrarian reform, Commissioners confronted almost insurmountable obstacles to policy implementation. The acute shortage of financial resources, the lack of technically qualified personnel (especially land surveyors and agricultural extension agents) and what Clark has referred to as 'chaotic' conditions in the countryside will be discussed later. It is clear that, throughout the deliberations, participants were fully aware of the wide gap between good intentions and the practical reality of the situation. García (1970, p.339) was less generous to the Commissioners and politicians: 'The theoretical superstition that it was enough to break the *latifundio* structure for the spontaneous and immediate achievement of a market economy, prevented the agrarian reform, during its first stage, from even studying the problems of agricultural marketing ...'.

It was widely accepted that any Bolivian agrarian reform policy would have to start virtually 'from scratch'. Before the National Revolution there had been 'only sporadic and isolated attempts at agricultural research on local crops and livestock': machinery, improved seeds, chemical fertilizers and pesticides were not generally used. 'The market system of the *colono* families was made up of labor exchanges, barter and small cash transactions among the many peasant families in markets located primarily in the provincial and cantonal capitals' (Clark, 1970, pp.21-23). Whilst Villarroel's education decree of 1945 had required estate owners to provide elementary schooling for their *colonos'* children, very few had done so, whether from lethargy or a fear of losing a ready supply of free labour. 'In 1951, only 27 per cent of the children between seven and fourteen years of age entered the rural schools' (García, 1970, p.328). The abysmal level of rural education, reflected in the *campesinos'* inability to read (either in their own language or Spanish, the official language) the simplest documents, or to follow written agricultural advice and technical instructions, presented a veritable stumbling block to any form of innovation.

The Agrarian Reform Law stipulated in general terms that *campesino* communities were expected to look after whatever community schools had been provided by former landlords and to 'create' schools in communities lacking educational provision. The Education Reform Decree of 1955 was to focus on the issue of rural education; it emphasized the importance of providing education in all rural settlements, as a means of integrating the rural population into national life. Article 175 stated that a simplified version of the Agrarian Reform Law would be printed in Aymara, Quechua and Guaraní, 'in order that the mass of *campesinos* in all rural districts' could 'take serious account of the new regulations' which would 'affect them'. Whilst access to educational facilities would have a considerable influence on later generations, it is doubtful whether more than a handful of *campesinos* in most provinces could digest the contents of the new law in the 1950s.

The Agrarian Reform Law's references to agricultural extension services and rural credit facilities were thinly scattered and extremely vague. Article 138 specified that income obtained from the sale of excess crops from collective land was to be used to pay for the services of a technical assessor, in order to bring about 'a permanent transformation of cultivation methods'. Article 165 described two of the National Agrarian Reform Council's responsibilities as: 'the organization of improvement systems, cooperation and agricultural credit' and 'the organization of colonization systems, rational exploitation and agricultural mechanization'.

According to a decree issued in September 1954, the *Banco Agricola* (Agricultural Bank), established in 1942, was an autonomous state institution, 'created to improve the country's agriculture': it would be 'at the service of Agrarian Reform'. The Bank was vested with the authority to carry out a number of tasks and to make available credit facilities and agricultural extension services'. Its basic duties were: to organize rural credit for agricultural producers; to devise 'a simple agricultural credit scheme, which would eliminate the activities of intermediaries and directly benefit individual *campesinos*, cooperatives and syndicates'; to offer advice and technical assistance to farmers, enabling them to increase their crop output; to promote saving in *campesino* communities and, with government funding, to set up rural farm shops, selling seeds, fertilizers, 'chemical substances', machinery and implements, including ploughs .

Agrarian reform institutions

The penultimate section of the Agrarian Reform Law was largely concerned with designating responsibilities to the institutions authorized to administer the new legislation. The National Agrarian Reform Service (SNRA) was identified as an autonomous body whose membership was to comprise: the president of the republic (the supreme executive authority); the National Agrarian Reform Council (CNRA); agrarian judges, rural agrarian reform assemblies and rural inspectors. Whilst the CNRA would function as the supreme appeal court in cases of land adjudication, and its director would be accountable to the Minister of Peasant Affairs, the SNRA was charged with the arduous task of interpreting and transacting the Agrarian Reform Law. The institution's principal functions were 'to initiate expropriation decrees, distribute land to peasant families, and draw up and implement the legal aspect of title distribution to all landholders' (Clark, 1970, p.31). The time-consuming, labourious procedures involved in land expropriation and redistribution were to begin with the lodging of *demandas* with the SNRA and end with the president's personal signing of individual land title documents.

According to Luis Antezana, an ex-Director of the CNRA, the signing of the Agrarian Reform Decree in August 1953 produced 'a profound psychological transformation' in the mass of the Bolivian peasantry (1955, p.7): the despised *indio* was converted overnight into the *campesino*. Whether or not the allegation was exaggerated, unquestionably the implementation of agrarian reform in conjunction with the MNR's radical programme of social, economic and political reforms, paved the way for a comprehensive restructuring of rural life in the republic. Within a very short period of time, some 400,000 peasant families became independent freeholders, liberated from the gross injustices of the *hacienda* system. Whilst the revolutionary programme of reforms did not guarantee sudden change and economic development per se., by breaking the shackles of the *colonato* system, enabling the *campesino* to participate actively in the monetary economy, by bestowing on individuals *la dignidad de la persona*, by providing access to education and authorizing - if not, encouraging - political activity, it made mobility possible in almost all spheres of life. In the years following the National Revolution traditionally inwardly oriented agrarian communities were exposed to the full impact of unprecedented forces and obliged to come to terms with an unfamiliar outside world.

In the second part of this book, the complicated issues and practical difficulties associated with implementing land redistribution, the impact of the 1953 Agrarian Reform Law on agricultural productivity and on the countryside in general, together with *campesino* opinions on the changes set in motion by agrarian reform legislation, are considered at length in the context of the Lake Titicaca region.

5 *Ley INRA* (1996)

Bolivia's highly controversial *Ley del Servicio Nacional de Reforma Agraria* (National Agrarian Reform Service Law), universally known as *Ley INRA*, entered the statute books in October 1996. Whereas the 1953 Agrarian Reform Law had, of necessity, been drafted within 90 days, the provisions of the new law were debated interminably over many months, during which time countless amendments were suggested and numerous drafts, discarded.

The objectives of *Ley INRA* (denoting the National Agrarian Reform Institute i.e. the executive body of the SNRA) are set out in the first of its 87 articles as follows: to establish the organic structure and attributes of the SNRA and the land distribution system; to guarantee landownership rights; to create the *Superintendencia Agraria* (Agrarian Superintendency) and *Judicatura Agraria* (Agrarian Judicature) and to establish their procedures, additionally, to regulate the process of *saneamiento* (eradicating landholding irregularities). Whilst *Ley INRA's* title and clearly stated objectives suggest the new legislation is solely concerned with the restructuring of the SNRA and rationalizing land distribution, even a cursory glance at the documentation reveals this to be far from the truth. According to Pablo Solón, an agrarian analyst, 'we are not talking simply of the organic structure and attributes of the SNRA but facing a law that makes profound changes in agrarian legislation' (1997, p.7). Inevitably, *Ley INRA* is widely referred to as Bolivia's new, or second, Agrarian Reform Law. For reasons discussed later, some politicians, economists and agricultural experts, including Antezana, remain vehemently opposed to many of the new measures and have been outspoken in their condemnation of *Ley INRA*, describing it as Bolivia's *contrarreforma* (counter-reform).

Ley INRA actually began life as a draft proposal *Ley INTI* (National Land Institute Law) early in the presidency (1993-97) of Gonzalo Sánchez de Lozada, who had served as Minister of Planning and head of the government's economic team during Paz Estenssoro's final presidency (1985-89). Following a disruptive demonstration by coca growers in August 1994, government assurances were given that the proposals contained in the *INTI* draft would not be presented to parliament until they had been discussed in detail with leaders of the national unions representing *campesinos*, coca growers and lowland indigenous peoples viz: the CSUTCB (the Confederation of Bolivian Peasant Workers), the CSCB (the Confederation of Bolivian Colonizers) and CIDOB (the Indigenous Confederation of Eastern Bolivia). At a meeting early in 1995 these three bodies presented a draft *Ley INKA* (National Institute of Kollasuyo and Amazonia Law), as a joint counter-proposal to the government's *Ley INTI*. Whilst acknowledging the need for restructuring the SNRA, the unions strongly urged that the debate be temporarily suspended and the matter further considered at a later date during general discussions on a comprehensive agrarian reform modernization law.

Further negotiations, involving the participation of ministers, *campesino* and indigenous peoples' representatives, in addition to delegates from the Chamber of Agro-Industrialists (with a membership drawn largely from the Santa Cruz region), led to the compilation at the end of May 1996 of the first version of *Ley INRA*: incorporating many of the *INTI* and *INKA* recommendations, this draft won the general approval of all parties concerned. Not surprisingly, consternation, confusion (exacerbated by textual ambiguities) and accusations of breaking promises followed the government's publication on the first day of August of the radically altered 'preliminary' draft of *Ley INRA: Modificación del Servicio Nacional de Reforma Agraria*. Government hopes that it would be accepted unchallenged were dashed immediately. Whereas in August 1953 the crowds of *colonos* and *de facto ex-colonos* gathered in Ucureña had enthusiastically applauded the signing of the Agrarian Reform Decree, in August/ September 1996 *campesinos* marched in their thousands from all parts of the republic to protest about the modified *Ley INRA* proposals.

Co-ordinated and incited by both national union and local syndicate leaders, and fearing that their lands would be sequestrated or heavy land taxes imposed, *campesinos* readily joined forces with groups of coca growers, deeply concerned about government threats to eradicate coca cultivation, and with lowland indigenous peoples, demanding the titles to territorial lands promised by the government of Jaime Paz Zamora after their 1990 'Territory and Dignity' march. All participants, together with large estate owners, who had publicly protested in Santa Cruz against land expropriation and taxing proposals, were united in their opposition to the establishment of the Agrarian Superintendency. According to the August version of the law, this newly created institution would be empowered to control and regulate land use, to insist on environmentally sensitive land being taken out of agricultural production, to fix the market value of land for

70

adjudication purposes and set the price to be paid by *campesinos* for expropriated property.

The writer considered herself very fortunate to be researching in the country during the period of the protest march and for all but the last two days of what became known as 'the siege of La Paz'. For several weeks 'the land question' dominated Bolivian news. Agrarian problems and rural development issues in general were fiercely debated on television and even in city streets. National newspapers contained daily photographs of the marchers and numerous articles by 'experts', eager to criticize the failures and shortcomings of existing legislation and to give vent to their strong feelings about the landowners and speculators who had over the years gained access to vast areas of land as a result of the corrupt practices of politicians and SNRA mismanagement. Even more unforgettable was the arrival in La Paz on 26 September of groups of ill-clad, banner-bearing forest Indians, last seen on the afternoon of 27 August being blessed by the Bishop of Santa Cruz before setting out on their month-long, arduous trek through inhospitable environments to join their fellow demonstrators in La Paz.

The march and subsequent 'occupation' of La Paz compelled the government to reconsider the *Ley INRA* proposals. Whilst some sections of the August draft were slightly adjusted in favour of smallholders and the owners of medium-sized properties, the final version failed to satisfy a number of *campesino* demands. Clauses detailing the responsibilities of the Agrarian Superintendency remained in place. Many *campesinos* are still convinced that they could have extracted further concessions from the government had not the leaders of the lowland indigenous groups decided in the early stage of the protest to break rank and accept government offers to begin separate negotiations about titling territorial lands. Although the new law bears the SNRA title, it will always be known by Paceños and the mass of *campesinos* as Ley *INRA* - so deeply engrained in people's minds are the dramatic events and endless debates which took place in the six weeks preceding the signing of the new law.

Why was *Ley INRA* introduced?

In his book aptly entitled, *La Tierra Prometida* (The Promised Land), published in 1995, Solón identified four key questions at issue in the ongoing debate about land and agrarian reform. What are the country's main agrarian reform problems? What should be changed? How can these changes be brought about? What are the likely effects of such modifications? To these pertinent questions must be added the equally intriguing one - Why did the clamour for a new, or modified, agrarian reform law reach a climax in the mid-1990s and not before?

As previously noted (p.48), Paz Estenssoro had never expected the 1953 Agrarian Reform Law to provide a complete and durable solution to Bolivia's multifarious and complex agricultural problems. It was accepted by most of the Commissioners that agrarian reform legislation would undergo modification and modernization in response to changing social, political and economic conditions. Even between August 1953 and the end of 1960 as many as 22 of the points raised in the Agrarian Reform Law were elaborated in supreme decrees and new 'sub-laws' (Serrano Torrico, 1993). Subjects ranged widely, as is apparent from the following selected titles: Rubber plantations (1954); Land surveyors (1954); Authorization to transfer lands directly (1955): Mobile agrarian tribunals (1956); Security of the stock farm (1956); Expropriation of land for public usage (1956) and Agrarian personnel in departmental offices (1958).

The ousting of Paz Estenssoro by the Bolivian armed forces in November 1964 marked the beginning of Bolivia's 'military interregnum', the eventful period between the suspension of democratic rule and its restoration in 1982. Although one of the initial acts of the incoming president, General René Barrientos, 'was to declare ... unswerving support for Agrarian Reform and for further distribution of land titles' (Klein, 1992, p. 248), any major modification of agrarian reform legislation in favour of the *campesino* was from that time out of the question. In August 1971, 'threats of extending agrarian reform to the new zones of commercial agriculture', made by President Juan Torres (described by Klein as 'the most radical and left-leaning General ever to have governed Bolivia'), contributed in large measure to his downfall: alarmed landowners in the Santa Cruz region readily financed the violent coup led by the then Colonel Hugo Banzer Suárez, himself a well-established member of the Santa Cruz landholding elite.

Neolatifundismo

Unquestionably, the military regimes of the 1960s and 1970s were largely responsible for what is regarded by most agrarian analysts as one of the major, and most intractable, agrarian reform problems confronting Bolivia in recent years viz. *neolatifundismo* i.e. the re-emergence and consolidation of the unproductive *latifundio*. Whereas, by 1964 landlessness and *minifundismo* (the excessive fragmentation of small land plots) had gained recognition as the overriding agrarian problems in the densely populated Lake Titicaca region, parts of the Yungas and the Cochabamba valley, at the other extreme, the processes involved in restoring the *latifundio* system were about to gather momentum in the departments of Santa Cruz and Beni, the Chaco of Tarija and some northern parts of La Paz. The re-establishment of the *latifundio* 'reached its culmination' between 1970 and 1980 (Alanes, 1993).

Agrarian experts, such as Solón and Antezana, trace the roots of *neolatifundismo* back to the 1953 Agrarian Reform Law:

Without doubt, the reconstitution of the *latifundio* is the central problem of agrarian reform in Bolivia. Recomposition of the *latifundio*, not by reason of its dimensions, which in general do not exceed the limits established by law, but *latifundio* because, as is stated in Article 12 of the Agrarian Reform Law, 'it remains unexploited or is exploited deficiently'. The principal mistake of the Agrarian Reform Law is that it opens the gates for the reconcentration of lands, does not set up mechanisms for avoiding the emergence of new *latifundios* and does not fix clear parameters for proceeding with the expropriation of newly-formed *latifundios* (Solón, 1995, p. 26).

In *Juicio y Condena a la Ley INRA* (Judgement and Sentence on the INRA Law, 1997), Antezana agrees that the shortcomings of the 1953 Agrarian Reform Law made the rapid spread of *neolatifundismo* possible, if not inevitable: by perpetuating 'the feudal-colonial' landholding structure, the MNR paved the way for the revival of the *latifundio*. Likewise, the failure to destroy the traditional *hacienda*, together with the retention of the archaic *sayaña-comunidad* based peasant landholding system, denied progressive *campesinos* the opportunity to develop a capitalistic form of agriculture. Of the 11,000 *haciendas* and *latifundios* in existence in the early 1950s, only 1,442 were totally expropriated, whilst the remainder were converted into 'medium *haciendas*' i.e. medium-sized properties and *empresas* (Antezana, 1979). Whilst a number of inefficient, non-mechanized *hacendados* retained some, or all, of their land by a variety of means (discussed later in connection with the Lake Titicaca region), an unknown number of former *latifundio* owners acquired new estates during the period of military rule. Additionally, land speculators, only interested in holding land as a hedge against rampant inflation, were able to join the ranks of *latifundistas*. According to Solón (1995, p.15), the owners of small and medium-sized properties were, by 1992, cultivating approximately 1,100,000 of the 4 million hectares to which they were entitled; at the same time the owners of large estates, including commercial farmers, were exploiting not more than 200,000 hectares of a total 19 million hectares considered suitable for arable farming purposes.

In a bulletin published in December 1992 and entitled, 'An Untenable Future? The Fate of Agrarian Reform', CEDOIN (Centre of Documentation and Information) summarized the features of 'the unproductive estate' as follows:

The agribusiness estates average 700 hectares ... Agribusiness interests cultivate only 0.3 per cent of the lands under their control, mostly for export, whilst vast tracts lie neglected and often unseen by their owners. In the tropical lowlands of Santa Cruz, the unintended underside of large scale, state sponsored "modern" agriculture is what experts call the "unproductive estate", marked by extensive rather than intensive production with relatively low yields, and accompanied by severe environmental destruction. The "unproductive estate" has thus reemerged following its intended abolition forty years before and today there is a greater concentration of uncultivated property than in 1953.

Neolatifundismo in the Bolivian Oriente is inevitably linked in the minds of most Bolivians with corruption and manipulation at the highest level. The 1953 Agrarian Reform Law had vested the president with 'the supreme authority' for granting land titles - a privilege that was frequently abused during the period of military rule. 'The President of the Republic transformed himself into a sort of Agrarian Reform God. As "supreme authority" he could not be questioned and he made the final decision on everything concerning the application of agrarian reform' (Solón, 1995, p. 18). No president used his land titling powers to greater advantage than Banzer Suárez.

Whereas 6 million hectares of land had been titled between August 1953 and 1964, and a further 4 million during the presidency of Barrientos, during his seven years in office, Banzer Suárez authorized the granting of 17 million hectares of land: 'the bulk of the lands titled came into the possession of enterprises and new *latifundistas*' (Solón, 1995, p. 22). According to a 1975 decree law on 'Issues relating to agrarian property', the *latifundio* remained 'illegal' and 'unrecognized': 'the state, through its specialist institutions' would 'dissolve any newly created *latifundio* splitting it up into *empresas*'. This was the process applied on countless occasions by the president, with the collusion of the SNRA: land grants of 10,000 hectares would be titled in the names of five individuals i.e. each recipient was granted 2,000 hectares, the legal maximum size for the *empresa agricola* in the Santa Cruz region. In 1993 Paz Zamora, in one of a number of such cases, signed annulment orders for the restitution to the state of 169,300 hectares of land distributed illegally. Significantly, the names of Banzer and Suárez appear eight times in connection with 18 land titles fraudulently granted in the department of Santa Cruz.

Although all presidents in office during 'the military interregnum' openly pledged their support for agrarian reform it is clear that, with the exception of Torres, their sympathies lay almost entirely with the landowning and business elites. Understandably, a number of agrarian experts apply the term of counter-agrarian reform to the period in question: indeed Antezana is not alone in insisting that this period of counter reform has never ended. 'All the counter-revolutionary agrarian legislation passed since the 1960s was destined to consolidate the *ex-haciendas* and to restrict the *campesinos*' property rights' (Antezana, 1997, p. 49).

One of the positive outcomes of the repressive regime of Banzer Suárez was the emergence in the mid-1970s of the Tupac Katari Indian-rights movement, whose leaders 'arose among the traditional rural *sindicatos*, especially among the previously more quiescent Aymaras' (Klein, 1992, p. 273). Following the return to democratic rule, the CSUTCB, established at the end of the 1970s by Tupac Katari officials, presented formal proposals to the government for making radical changes to the existing agrarian reform legislation. As Paulino Guarachi, executive secretary of the CSUTCB, commented in 1992: 'Since 1983' (a year of severe drought and very low agricultural output in much of the republic) 'the CSUTCB has been advocating a new Fundamental Agrarian Law based on the idea that 'the land belongs to those who work it'. And this would mean directly, personally working the land - not in the sense of businessmen who 'manage land that other people work for them'. Guarachi went on to say that:

> The CSUTCB presented the Fundamental Agrarian Law as a means of strengthening the peasant and indigenous community, its authorities, and its capacity to produce and market crops, to obtain credit, and to take advantage of development opportunities. But this Law was rejected by Congress (CEDOIN, 1992).

Amazingly, with minimal assistance, the confederation had drafted an almost comprehensive agrarian reform law on behalf of its *campesino* members.

When Paz Estenssoro took office for the fourth and last time in 1985 *campesinos* entertained high hopes of achieving their agrarian reform goals. The government's draft General Law of Agrarian Development (1988) placed 'the emphasis on low productivity as the principal cause of the backwardness and poverty of *campesinos* and, in consequence, established as its objective the need to procure a greater efficiency in productive activity, especially of small producers' (Solón, 1995, p.26). Whatever the intentions, Paz Estenssoro's New Economic Plan, a radical structural adjustment programme introduced in 1985, to tackle the problems of 'a bankrupt public sector and a deeply depressed private economy' (Klein, 1992, p. 271), worsened by the collapse of the tin market and runaway inflation (inflation exceeded 20,000 per cent in April 1985), proved disastrous to small farmers. The subsequent bankruptcy and closure of the Agricultural Bank deprived rural communities and individual *campesino* families of any means of obtaining outside credit for agricultural purposes: additionally financial support was withdrawn from government agricultural research institutions and agricultural extension agents ceased to operate. Instead of paving the way for 'modernizing' the 1953 Agrarian Reform Law, the New Economic Plan resulted in the withdrawal of virtually all government support for subsistence farming.

Whilst the government of Paz Zamora also viewed low agricultural productivity as the country's main agrarian problem, it demonstrated little respect or sympathy for the *campesino* population: the New Agrarian Law proposal of 1989 affirmed that 'in Bolivia there is no unproductive *latifundio*, that the *minifundio* or small *campesino* property is inert land (without life) and has no economic significance...that legal title to private agricultural property is obtainable when the land becomes economically productive' (Solón, 1995, p. 27). No wonder that Guarachi felt justified in concluding that:

> The neoliberal vision of land distribution is based on the concept that 'the land belongs to those who have the money to buy it' and that the land is an object like any other commodity which you can buy and sell. We think that to consider land an object is irrational - it may make sense to other people but not to us... Is land just something to exploit for profit? The Aymara, Quechua, Guaraní cultures would say just the opposite. We live together with the land. There is complementarity between people and the land (CEDOIN, 1992).

Without any doubt the lowland indigenous peoples' march for 'Territory and Dignity' added fresh impetus to the struggle for a more equitable distribution of land and recognition of territorial rights. Until the lowland indigenous groups began to organize in the 1980s, 'they did not exist as far as the rest of the country and world were concerned. They did not identify themselves with the class struggle of the *campesino* and trade union movement but fought for their rights and identity as original communities' (CEDOIN, 1996). For the 35 indigenous groups in Amazonia and the Chaco, in the departments of Santa Cruz, Beni and Pando, land and territory have always been vital issues.

> When we talk of "land" we mean the consolidation of community lands. We mean the legalising with titles the land where we have been living from time immemorial. We also use the term "territory" because we are looking for territory for the Chiquitano people. "Territory" also means the right to use, manage and administer all the natural resources of our territory - both above and under the ground, the flora, fauna, water and air of the geographic space which we occupy as legitimate owners (the comments of a Chiquitano leader quoted by CEDOIN, 1996).

Whereas in former times tribal groups had to contend with encroachment on territorial lands by rubber tappers and gatherers of Brazil nuts, over recent decades, lands traditionally occupied by such groups, but to which they have had no legal titles, have been invaded by 'economically and politically powerful agroindustrialists, cattle-ranchers, logging companies, fishermen, poachers and large commercial enterprises' (CEDOIN, 1996): natural gas and oil prospectors have also caused havoc and consternation. In September 1990 the central government was taken by surprise when, for the first time, several lowland indigenous groups, co-ordinated by the leaders of local organizations, decided to challenge head-on the legality of the government's granting of extensive logging concessions in the Chimane Forest region of Beni. More than 800 Chimanes,

Moxeños, Yuracarés and Guaranís, intent on marching from the Beni capital of Trinidad to La Paz (a distance of 700 km), were joined *en route* by significant numbers of supportive Quechua and Aymara *campesinos*. The historic nature of the event was indicated by the president's decision to welcome the marchers before their arrival in the city. After a week of negotiations, the protesters were given an assurance that all the mahogany loggers would be removed from the area in question within three months. Additionally, Supreme Decree 22610, issued on 24 September, recognized the Isiboro-Sécure National Park as an 'indigenous territory'.

Whilst leading directly to close cooperation between *campesino* unions and indigenous groups' organizations, the 1990 march also set the tone for demonstrations two years later marking the 500th anniversary of the so-called 'discovery of the Americas'. According to Esteban Calani Gonzáles, one of the founders of the CSUTCB, '65,000 *originarios*' held protests in La Paz on 12 October 1992, in recognition of '500 years of resistance', whilst at least 100,000 others demonstrated elsewhere in Bolivia (1996, p.176). The events of the following month were to prove equally, if not more, dramatic.

Less than six weeks after the Columbus 'celebrations', in response to the rising tide of mismanagement and corruption allegations against the CNRA, the government deliberately released the comments on Bolivian agriculture contained in a World Bank report. The document focused attention on the constraints confronting the mass of *minifundistas* viz. excessive land fragmentation, accelerating land degradation as a result of soil erosion and salinization, the lack of rural infrastructure and of opportunities for *campesinos* to increase agricultural productivity and improve their living conditions. Whilst the report was seen to be highly critical of the country's agrarian reform policy in general, it was particularly scathing about the CNRA, referring to it as being 'extremely disorganized and bureaucratic'. Almost immediately afterwards President Paz Zamora took the drastic and unprecedented step (described by him as 'an historic measure') of completely closing both the CNRA and INC (the National Colonization Institute). Although the CNRA had been criticized over the years for a number of shortcomings, particularly its excessive bureaucracy and the lengthy procedures and delays involved in titling *campesino* lands, its final downfall was precipitated by the infamous Bolibras scandal, the most notorious of numerous land scandals in the Santa Cruz region.

Shortly before the closure of the CNRA, Edim Céspedes, Minister of Education and Culture, had been made to resign from Congress, accused of having abused his position as a minister of state by illegally acquiring titles to at least 95,000 hectares of land in the department of Santa Cruz; this he had done as president of Bolibras, which claimed to be an agribusiness but operated primarily as a land speculation enterprise. Whereas both the CNRA and INC had proclaimed that no more land was available for distribution within the area - and many Bolivian *campesinos* had had to wait at least 15 years to secure their land titles - Céspedes had been able to obtain land titles in record time: his 'enormous land grant was surveyed in just one visit, and the agrarian reform judge in Santa Cruz approved the adjudication in under 50 days' (CEDOIN, 1992). The degree of corruption and 'influence-peddling' involved in the Bolibras case forced the president to exercise his supreme authority to annul the land grants. The subsequent suspension of all CNRA activities made modifications to the existing agrarian reform legislation inevitable and urgent. According to CEDOIN it opened 'the door for neoliberal restructuring of the country's land tenure system'.

Sánchez de Lozada, the MNR leader of the 1993-97 coalition government, by reason of being a prominent mining capitalist, was as only to be expected identified with the big business elite; during his presidential campaign he had declared his total commitment to neoliberal economic policies. Nevertheless, until August 1996, many *campesinos* were inclined to believe that some, if not all, of their hopes and demands would be satisfied by his promised agrarian reform legislation. His vice president, Victor Hugo Cárdenas, the first Bolivian Aymara to secure high political office - appointed to attract the Andean *campesino* vote and as a symbolic gesture - had himself been born in the farming and fishing community of Huatajata, beside Lake Titicaca: he had frequently stated publicly that he had the *campesinos'* interests at heart. Moreover the president had fulfilled his promise to introduce an education reform law and one aimed at decentralizing power - 'to incorporate the indigenous, *campesino* and urban communities in the legal, political and economic life of the country': according to the president, the structural changes effected by his 1994 Law of Popular Participation constituted 'the most important redistribution of political and economic power in the Republic since the 1952 Revolution' (CEDOIN, 1995).

Throughout the lengthy period of discussions on *Ley INRA* two principal objectives guided the government's thinking and planning. Firstly, it was considered imperative to regularize as quickly as possible the processes of land titling, expropriation, the reversion of confiscated lands to the state and their subsequent redistribution, whilst at the same time ensuring that the administrative structure established to replace the dismantled CNRA would operate with greater efficiency and speed, avoiding all forms of corruption and manipulative practices guaranteed to lead to the sort of 'land scandals' which had become all too familiar since the 1970s.

The second objective, inextricably linked with the first, was to increase agricultural productivity in both the commercial and *campesino* sectors. There was agreement amongst government parties that all large landholdings should be expected to yield an 'economic benefit' to the state, in the form of agricultural output and land tax payments. The unproductive *latifundios*, claiming to be agricultural enterprises, and estate lands illegally acquired by land speculators, should be subject to expropriation and returned to state ownership, thereby creating a land market: it was later argued that more leniency should be applied to well farmed properties obtained through the use of influence and by other devious means. *Campesinos* should be provided with adequate, unspecified 'support', to enable them to become efficient cash crop producers.

Increased agricultural productivity?

Paz Estenssoro's preamble to the 1953 Agrarian Reform Law had stressed the need 'to raise the country's actual production level'. Although the first three of the six 'fundamental objectives' had been largely concerned with 'the liberation of the *campesinos*', they had clearly indicated that the revolutionary government expected untold economic benefits to be derived from the abolition of the inefficient *latifundio* system; the second objective referred directly to 'the modernization' of peasant agriculture. Doubtless, a number of Commissioners had regarded the fourth objective - 'to stimulate greater productivity and commercialization of the agricultural industry' - as the main priority of agrarian reform. To what extent had this objective been achieved?

Carter, researching on the Altiplano since pre-1953 agrarian reform times, was to recall the dispossessed landlords' delight 'in arguing that the reform was accompanied by a severe decline in agricultural production', demonstrated by 'the grave food shortages suffered by the cities during the mid-1950s' (1971a, p. 68). He attributed such temporary shortages mainly to 'the change in the system of distribution' i.e. to the inevitable marketing readjustments once the ex-landlords' 'virtual monopoly on agricultural wholesaling had been broken'. Carter's observations led him to comment enthusiastically on the state of agricultural productivity in the early 1970s: 'except for wheat, supplies of foodstuffs flowing through the urban market place over the years have returned to normal and, in many cases, have noticeably increased'. City consumers had 'benefited' financially since, unlike former *hacendados, campesinos* were unable to store commodities in bulk until sale prices increased.

Burke, also reviewing his research findings in the early 1970s, whilst acknowledging the difficulties involved in comparing 'the economic performance of the pre-reform Bolivian *haciendas* with the present-day *ex-haciendas*, because comparisons over time may reflect climatic or price changes above all else', agreed with Carter that 'total production as well as market production exceeded pre-reform levels'. From his mid-1960s survey of four *ex-haciendas* in the Lake Titicaca region he drew the following general conclusions about agricultural productivity: 'The evidence accumulated in this study suggests that labor productivity on the Bolivian *ex-haciendas* has decreased, land productivity has increased, and capital productivity has remained unchanged since the land reform'. This situation he attributed in large measure to 'the increased population, greater use of the more marginal land, and the small decrease in agricultural equipment', adding that 'in part, the decrease in labor productivity reflects the increase in leisure and off-the-farm employment of the Bolivian *campesinos*' (1971, p.316).

Most other researchers have been reluctant to make categorical claims about agrarian reform's impact on agricultural productivity - whether in terms of output per hectare, worker total production or agricultural gross domestic product. Whilst there is no dispute about dramatic increases in the production of commercial export crops (such as cotton, sugar, rice and soya) on large estates in the Oriente, the changing output and value of food crops for national consumption in traditional farming areas have, for a number of reasons, proved virtually impossible to quantify. Clark (1970, p.62) outlined the almost insuperable problems in evaluating economic efficiency. 'It is difficult to determine what has happened in the productivity of either land or labor since the land reform in Bolivia; there are no benchmark statistics which can be compared with more recent studies ... too many changes have taken place in opposite directions to arrive at a clear conclusion'. Likewise, there were 'no quantitative data' which could 'show increases or decreases in income to land reform beneficiaries'.

The passage of time has not clarified the situation. As in many other Latin American republics, reliable statistical data on *campesino* production and marketing transactions are virtually non-existent; in any case a significant proportion of the food produced is still bartered in local markets for other types of food or essential commodities. Thiesenhusen (1995, p.64) notes that 'the effects of reform on agricultural production has been a hotly debated issue in Bolivia'. After considering a variety of factors (including post-reform increases in *campesino* food consumption, land 'idled' in the years immediately after the introduction of agrarian reform, extreme population pressure on cultivable land resources in some areas, government imposed underpricing of farm produce in urban markets, 'inadequate applied research combined with ineffective extension services', Bolivia's unfavourable trade balance 'from 1951 to the mid-1970s' and the country's depressed economy in the 1980s), Thiesenhusen draws a very general, brief conclusion. 'Initially land reform brought peasant beneficiaries a higher level of living, which stabilized when credit and inputs to increase productivity were not forthcoming; eventually inflation eroded the gain'.

Despite the inadequacies of, and difficulties in, interpreting available statistical data, it is clearly apparent that the output of staple foods from small and medium-sized holdings in parts of the republic has increased significantly over the years since 1953. Bolivia's agricultural system was producing only 40 per cent of the food required by its population of 3,109,131 in 1950, according to that year's national census returns i.e. at a time when peasant communities were subsisting on the meagre products from *sayaña* plots, the country's *hacendados*, in possession of some 31 million hectares of agricultural land, were failing miserably to meet the consumer demands of the urban sector. By contrast, today's *campesinos* and owners of medium-sized properties produce approximately 60 per cent of the food consumed by a population of 8 million; roughly 20 per cent is produced on large estates and the remainder, imported, some of it in the form of donations.

Bolivia's agricultural productivity and export earnings are subject to violent fluctuations. Severe weather conditions, especially prolonged droughts, affect commercial farmers and *campesinos* alike. In the early months of 1996 eastern lowland stock farmers claimed to have lost US$ 70 million in earnings from milk and beef because all pasture land had been destroyed by drought; additionally, sorghum, maize, wheat and sunflower harvests proved disastrous. *El Niño* droughts caused considerable hardship to *campesino* communities the following year. According to the Ministry of Agriculture, the national potato harvest was 40 per cent less than normal, leading to a loss of revenue from potatoes alone of approximately US$ 79.5 million. In the department of La Paz onion production was reduced by 39 per cent, wheat by 37 per cent and potatoes by 29 per cent. Bolivia's agricultural export earnings for the first nine months of 1998 were 24 per cent less than for the same period in 1997 (*Latin America Weekly Report*), partly because of further droughts, but also because of increasing competition and the fall in world prices for certain products, especially soya.

Controversial issues

The proposed *Ley INRA* was an anathema to the *empresarios* (commercial farmers) and *latifundistas* of the eastern lowland regions. They were well aware that they had nothing to gain from it; indeed, those whose properties had been acquired through corruption and guile knew they had everything to lose. All landowners were adamant that INRA should not be given the authority to deprive farmers of cultivable land in order to make way for 'conservation activities, protection of biodiversity, research and ecotourism'. Likewise there was strong opposition to the proposed rise in land taxes; payments were to range from 0.3 per cent of the land value in the case of medium-sized properties to 1.5 per cent for the largest estates. The proposition that non-payment of the annual tax - especially after the sort of drought-caused disastrous harvest experienced that year - could result in land being expropriated, was strongly condemned.

The announcement on 7 October 1996 that the proposed *Ley INRA* had been finally approved and sanctioned by the government led to an extraordinary sequence of events in Santa Cruz. Land owners took possession of both the prefecture and town hall and called for a 48 hours general strike. They then proceeded to distribute chickens, eggs, milk, bananas and vegetables to local residents and *campesinos*, as they said, to symbolize the income they would stand to lose once the new land tax was imposed. At a public meeting they declared the president to be *persona non grata.*

Unlike the commercial farmers, *latifundistas* and the mass of *campesinos,* the indigenous peoples, led by Marcial Fabricano, the executive secretary of CIDOB, were convinced that they had nothing to fear from any new law since they had no rights or privileges to start with. With the arrival of the Spaniards they had lost - and never regained - their rights by possession to territorial lands. Although the vice president had been holding regular discussions with their representatives, very few of the supreme decree promises made after their 1990 march to La Paz had been fulfilled. Once more they were having to undertake the gruelling march to the city to plead for legal titles to territorial lands occupied by their ancestors from time immemorial. Each tribal group was claiming titles to specific geographical areas e.g. the Chiquitano people were laying claims to 1,440,000 hectares of land in total, within three regions. The Chiquitano leader summed up the group's difficulties in preventing incursions: 'we are up against magnates with economic and political power ... The issue of land titles is vital because these powerful companies are saying we have no legal documents and are not the owners of the land' (CEDOIN, 1996).

The indigenous peoples asked for assurances that there would be no further encroachment on territorial lands by speculators or colonizers. If timber companies were going to be granted concessions in neighbouring areas, they should be completely banned from entering adjacent titled land. Apart from causing environmental devastation (it was claimed that for every mahogany trunk removed, 70 other trees were destroyed), loggers were hunting animals recklessly and their cutting equipment was scaring other animals away. Indigenous peoples should be given exclusive use of all resources within their titled land areas and their 'intellectual property', protected. Unscrupulous researchers and pharmaceutical companies should be prevented at all costs from removing indigenous plants without adequately reimbursing the local population.

Shortly after arriving in La Paz, the leaders of the indigenous peoples' contingent agreed to accept the government's overtures to negotiate separately and withdraw from the demonstration, an action fully understood by *campesino* protesters but, nevertheless, strongly resented.

In the early stages of the mass protest, leaders of *campesino* and indigenous unions (including CSUTCB and CIDOB) met together in Santa Cruz, as a 'technical commission', to reach a consensus on a list of agrarian reform demands to be presented to the government: the final document contained 76 articles in total. At the outset, the government was accused of abandoning former promises; government attempts to reassure *campesinos* that the new Superintendency would secure their land ownership rights were dismissed as a meaningless ploy to persuade the protesters to disperse. Cárdenas was later to admit that the *campesinos'* confusion was partly attributable to the government's failure to explain in detail the reasons for amending the *Ley INRA* proposals agreed in May; no real attempt had been made to answer *campesino* queries, thereby allaying their fears.

In the opening section of the document, union leaders reaffirmed their approval of the government's May draft *Ley INRA* proposals and demanded the withdrawal of the amended version. Failing that, all clauses referring to the detested Superintendency should be deleted: its establishment was 'unconstitutional' and would never meet with *campesino* approval. In view of the countless abuses of political power and influence in the past, no future president should be allowed to make the final decisions on whether to grant or withhold land titles. *Campesinos* were opposed to the incorporation of the Ministry of Sustainable Development and the Environment within the structure of the SNRA: they objected to the clause giving the minister authority to expropriate property 'for the protection and conservation of biodiversity'. *Cocaleros* were particularly concerned about various passages in the government's draft proposals referring to the ministry's role of approving 'regional and national plans for the use of the soil' and the hint that the minister would be able to regulate precisely which crops should be grown in, and which excluded from, any particular area. The growers of coca for cocaine demanded that if the government insisted on eradicating coca, they should receive realistic advice on alternative, replacement crops i.e. crops marketable at prices comparable with those obtained from selling coca (it was claimed that coca yielded at least five times the income derived from the cultivation of any other known crop).

There was general acceptance of the urgent need to expropriate *latifundio* lands and the properties of non-farming speculators. In isolated parts of the republic, where *ex-hacendados* had been able to reassert their control over their former estates and their authority over *ex-colonos*, *campesinos* welcomed the prospect of reclaiming their plots of land through the process of *saneamiento*. On the other hand, the adjudication of unworked or contentious *campesino* lands should only be carried out at the request of the *campesinos* themselves and under no circumstances should fields worked by *campesinos* be subject to expropriation. Colonizers in the eastern lowlands maintained that if they had occupied land for at least two years before the promulgation of the new law, they should be allowed to retain it without payment. Foreigners should not be granted land titles anywhere in the country (the proposed *Ley INRA* stipulated that the ban be restricted to land grants within 50 km of national boundaries).

One of the main demands made by *campesinos* was full recognition by the government of land **ownership,** not merely **working** rights, as had previously been the case. They argued that the slogan, 'the land belongs to those who work it', had been quoted *ad infinitum* since 1953 but, in actual fact, *campesinos* had been denied outright ownership of the soil - and subsoil - by successive governments. **All** small farmers should be exempt from paying any form of land tax. In the August version of *Ley INRA*, the term 'adjudicate' had been replaced by 'sell' expropriated lands. *Campesinos* demanded an assurance that they would not be pressurized into selling land, or be in danger of having land seized for non-payment of debts. Whilst some wanted to prohibit the sale of all *campesino* land, this was certainly not a feeling shared by all. For example, a number of *campesinos* in the study area had already sold plots of land bordering Lake Titicaca to restaurateurs, speculators etc. before migrating to La Paz or El Alto: some rural-urban migrants had sold - or were waiting to sell - inherited tracts of land for the same purposes. Such individuals wanted a free market in land i.e. to be able to negotiate sale prices in an open market, rather than have land prices fixed by the Superintendency, regardless of soil quality and location.

Throughout *campesino* discussions the main focus was on land and territory - as it was in the proposed *Ley INRA*. What was widely referred to as an agrarian reform law dealt almost exclusively with land reform issues. Yet for many *campesinos* living in densely populated regions, it was abundantly clear from the start that the process of *saneamiento* was unlikely to benefit them in any way. *Campesinos* interviewed in the Lake Titicaca region in August 1996 recognized that there was no solution to the acute problems of *minifundismo* and landlessness, unless they were prepared to move to colonization zones or decided to abandon farming and migrate to urban centres. The expressed needs of such *campesinos* ranged widely but were primarily concerned with the means of increasing agricultural output from small 'fixed' farm units, rather than expanding the area of land under cultivation. Amongst other agrarian reform problems, they specified the withdrawal of farm extension services, lack of access to research centres and difficulties of obtaining rural credit for the purchase of animals, medicines, seeds, tools, fertilizers, pesticides etc. They also wanted guaranteed minimum prices for produce, compensation for crop losses as a result of severe weather, and restrictions on importing food donations, undermining pricing mechanisms in city markets.

The provisions of *Ley INRA*

Although *Ley INRA 's* regulations are presented within a complicated framework of broad headings, chapters, sections and articles, they fall naturally into five main components. The preliminary section, entitled 'General Dispositions', and incorporating Articles 1 to 4, is followed by a number of chapters, together containing Articles 5 to 40, concerned exclusively with the restructuring and functioning of the SNRA. Articles 41 to 50 refer to different categories of agricultural properties and lay down regulations relating to land ownership and distribution; Articles 51 to 87 justify the land reform processes of *reversión* (reversion of land), *expropiación* (expropriation) and *saneamiento*, and outline the procedures to be adopted in order to achieve a rational land redistribution. The text concludes with 25 wide-ranging 'Transitory' and 'Final Dispositions'.

Having established the main objectives of *Ley INRA* (see p.69) the introductory section sets out a number of guiding principles. At the outset a sharp distinction is drawn between 'the functions' of commercial estates and medium-sized properties on the one hand, and all forms of *campesino* and community holdings on the other. 'The *campesino* residential/ground plot, the small property, the community property and the lands of original communities fulfil **a social function** when they are destined to bring about family well-being', whilst the commercial estate and medium-sized property serve 'a **socio-economic function**'. The **socio-economic function** 'in agrarian terms ... is the sustainable usage of land in the development of agriculture, forestry and other forms of productivity, as in conservation and the protection of biodiversity, research and ecotourism ... to the benefit of society, the collective interest and that of the owner'. As agrarian analysts, such as Antezana, contend, this type of negative analysis (implying that peasant farmers are tied to subsistence farming and serve no economic function), together with the continued emphasis on community - rather than individual - farming and the general lack of government financial and advisory support, is guaranteed to do nothing to advance the prospects of progressive, capitalist-minded young *campesinos*.

Article 3, whilst guaranteeing the landownership rights of *campesinos* and indigenous peoples, and asserting that commercial estates and medium-sized properties 'enjoy the protection of the state, in as much as they perform a **socio-economic function** and are not abandoned', states categorically that the *latifundio* is illegal. It is interesting to note that *Ley INRA* adopts precisely the same wording as the 1953 Agrarian Reform Law i.e. *el estado no reconece el latifundio* (the state does not recognize the *latifundio*). The same clause takes Paz Estenssoro's earlier concern for equality further, by including a new and welcome anti-discrimination regulation: 'equity criteria will be applied in the distribution, administration, possession and exploitation of land in favour of the woman, regardless of her civil state'.

Whilst exonerating all *campesinos* and communities from land tax payment, Article 4 states that owners of medium-sized properties and large estates will be required to pay land tax 'in accordance with the value placed on the property by the owner' - a highly controversial provision discussed later.

The restructured SNRA

The SNRA is 'the organism responsible for planning, executing and consolidating the process of agrarian reform in the country' (Article 5). Under the new law, the SNRA has four constituent members: 'the President of the Republic, the Ministry of Sustainable Development and the Environment, the National Agrarian Commission and the National Agrarian Reform Institute' (INRA). The president retains supreme executive authority within the SNRA restructuring, but is relieved of the former judicial powers. His role is 'to consider, approve and supervise the formulation, execution and fulfilment of land distribution, regrouping and redistribution policies'; to appoint or nominate agrarian reform departmental heads i.e. the Superintendent, the Director of INRA and the Minister of Sustainable Development, and to grant land titles, or authorize departmental prefects to do so on his behalf - a point disputed by *campesinos*, because of the possibility of 'political manipulation' (Solón, 1997), The president is also expected to issue supreme resolutions in situations calling for immediate action e.g. a 'land scandal' such as Bolibras.

The powers of the Minister of Sustainable Development to classify land according to its use and 'to urge the expropriation of land for conservation and protection of the environment' have already been noted (p.83): in its agrarian reform role, the Ministry is also required 'to evaluate and plan the use of land resources and the application of appropriate technologies'. The National Agrarian Reform Commission (CAN) is 'the organ responsible for planning and proposing land distribution, regrouping and redistribution policies and, whatever their condition or use, to bring them to the notice of the supreme authority of the SNRA'. The Commission acts as a consultative body and has a number of attributes, including the following: the control and supervision of the procedures involved in carrying out land reform; exercising 'social control over abandoned land' and land not serving a socio-economic function, and safeguarding the rights of indigenous peoples. The Commission has eight members: four of these are members of the government, one represents commercial farmers and the remaining three are *campesino* and indigenous peoples' delegates nominated by the CSUTCB, CSCB and CIDOB. According to Article 16, departmental commissions are to perform similar functions (at regional level) to the national institution; they are required to monitor *saneamiento* procedures and 'to channel petitions' to higher authorities.

The newly created INRA corresponds to the former CNRA but is not vested with the same legal powers and does not have the same close involvement with *campesino* syndicates. It is 'the technical-executive organ, charged with directing, co-ordinating and activating the policies laid down by the SNRA' (Article 17): according to Article 18, INRA is responsible for supervising the procedures of reversion, expropriation and *saneamiento*, determining which lands should be granted outright and which should be subject to adjudication. It also has a duty to try to resolve landownership conflicts and to collaborate with local public and private bodies 'in the provision of basic services and technical assistance'.

Article 26 specifies the numerous tasks allocated to the feared Superintendency. Apart from those referred to previously (p.70), the Superintendency is to classify lands 'according to their best usage'; to advise on the expropriation of lands, by reason of 'their failure to perform a socio-economic function'; to keep registers of land use and soil potential; to inspect lands and institute 'the precautions needed to avoid land being exploited in a form contrary to its capacity and best usage'. Agreement was reached prior to the law's enactment that the first Superintendent should be appointed for a period of 10 years: thereafter the period of office would be five years. Agrarian justice is administered by the Agrarian Judicature, which 'is independent in the exercise of its functions and subject only to the Political Constitution of the State and its law' (Article 31).

According to Cárdenas, vice president at the time of *Ley INRA 's* enactment, 'the managers of INRA need to be people who are professionally capable, politically independent and equidistant between businesses, indigenous communities, *campesinos* and foreign landowners' (*BT,* 1997). Whilst *Ley INRA* has been designed to distribute the powers and responsibilities previously enjoyed and exercised by the president amongst a number of agrarian reform institutional heads, it is virtually impossible to eliminate the threat of bias and political influence and corruption. This became clearly apparent in November 1997, when the recently inaugurated president, Banzer Suárez, himself responsible for so many land titling irregularities in the 1970s, appointed Dr. Hugo Teodovich as Director of INRA. Within the first week Teodovich had set out proposals to modify *Ley INRA* radically i.e. to reduce, or abolish, land tax for businesses and owners of large properties, to introduce legal guarantees that private property could not be reclaimed by the state and to expropriate *campesino* land for non-payment of bank debts.

Agrarian property and land distribution

Like the 1953 *Ley de Reforma Agraria, Ley INRA* recognizes six forms of agrarian land holdings, the first four of these are described in terms very similar to those used in the original legislation (pp.52-53):

1 *El solar campesino* (residential/ground plot) 'constitutes the place of residence of the *campesino* and his family'. It can not be sub-divided and is not subject to land taxation but can be sold in accordance with the law Such a holding cannot be sequestrated unless it is required by the state for 'public utility' purposes.

2 *La pequeña propiedad* (the smallholding) 'is the source of subsistence resources for the owner and his family'. The same conditions apply as in (1).

3 *La mediana propiedad* (the medium-sized landholding), employing 'salaried workers and using mechanical equipment', produces mainly for the market. The property can be sold, pledged or mortgaged; the owner must pay a tax on the land.

4 *La empresa agropecuaria* (agricultural enterprise, replacing *la empresa agricola*) 'is exploited with supplementary capital, a salaried work regime and the use of modern techniques'. It is subject to the same regulations as (3).

5 *Las tierras comunitarias de origen* (indigenous community lands, roughly equivalent to the 1953 category of *la propiedad de comunidad indígena*) are 'geographic spaces that constitute the habitat of indigenous peoples and original communities, to which they have traditionally had access and where they maintain and develop their own forms of economic, social and cultural organization, in a way that ensures their survival and development'. Such lands can not be broken up, mortgaged or sold and are not subject to taxation.

6 *Las propiedades comunarias* (replacing *la propiedad agraria cooperativa*) are properties that are titled collectively to *campesino* communities and *ex-haciendas* and constitute the resource base for the subsistence farming of their owners : they are subject to the same conditions as (5).

Article 41 ends with a somewhat vague reference to the legally permitted size of different types of landholding: 'the extensions of agrarian properties ... will be regulated' taking into account 'agro-ecological zones' and regional variations in potential agricultural productivity. Whilst the 1953 clauses on property dimensions are amongst those listed in *Ley INRA* as still being valid, uncertainties about future changes are causing considerable concern to *campesino* unions and agrarian analysts alike. Solón (1997, p. 13) sees two possible dangers. If no limitation is placed on the size of large estates, there will be nothing to prevent owners from endeavouring, by whatever means, to push property boundaries way beyond former legal limits. On the other hand, there is the fear of *campesino* owners of small properties being obliged to pay land tax if regulations on the maximum size of property are standardized nationally. For example, it could be decided to restrict the size of the small property to 10 hectares (the maximum size allowed by the 1953 Agrarian Reform Law in the Lake Titicaca region and much of the Yungas), forcing smallholders from the southern Altiplano (where the maximum size was previously set at 35 hectares, to compensate for the region's harsh climate and poor soils) to pay land tax, as re-classified medium-sized property owners.

Article 42 discriminates further against individual *campesinos* intent on expanding their cash-cropping activities. *Ex-hacienda* and community lands made available for distribution are to be granted 'exclusively' in favour of groups of indigenous peoples and *campesino* communities, 'represented by their leaders and *campesino* syndicates'. According to Article 46, foreign states are not entitled to acquire land grants anywhere inside the republic; foreigners who have resided and farmed in the country for 'some time' are eligible to obtain land away from the 50 km frontier zone, 'from a third party'. Public functionaries including the national president, senators, the Agrarian Superintendent, the Minister of Sustainable Development and the Environment and departmental prefects, are prohibited from procuring land titles whilst in office and during their first year after leaving government or SNRA employment.

Land redistribution

Much of the remaining text is concerned with regulations governing the state's retrieval and redistribution of land plots and estates through the processes of reversion, expropriation and *saneamiento*. Reversion is defined in Article 52 as the return of land to the public domain, without any payment of compensation or indemnity. According to *Ley INRA*, medium-sized properties and large estates should revert to state ownership if they are abandoned. However the payment of land tax 'is proof that such land has not been abandoned'. Thus a *latifundista* or speculator, who may never even have visited the property in question, can retain control of it if he is prepared to pay the land tax regulated at one per cent (rather than the previous two per cent) of the self-assessed value. As Solón remarks, *Ley INRA* makes the reversion of the *latifundio* almost impossible: indeed, the individual *latifundista* may be required to pay considerably less than he did before the enactment of the new law.

Expropriation, involving the payment of compensation, is intended to occur in situations where 'the land does not serve a socio-economic function' or in order 'to satisfy public utility needs' (Article 58). Whilst only medium-sized properties and large estates can be expropriated for the first reason, the owner of any type of property may be deprived of land when such an action is considered necessary to satisfy the interests of the country, region or town. In addition to expropriation of land for conservation purposes and the protection of biodiversity, this category also includes expropriation arising from the need to carry out 'works of public interest, such as the building of a road or a school'. Additionally, medium-sized properties and estates can be expropriated in order to regroup and redistribute lands to *campesino* communities and indigenous peoples, 'in accordance with socio-economic necessities and rural development'. Compensation to landholders is to be based on their self-assessed property values - a device believed by drafters of the bill to prevent landholders from under-valuing property in cases where there is even the slightest threat of expropriation. *Campesino* communities and indigenous peoples are to be paid the current market value, determined by the Agrarian Superintendent.

Ley INRA draws a clear distinction between 'relative nullity' and 'absolute nullity'. A landholder issued with a resolution of relative nullity for a minor infringement, such as falsifying statistical data relating to crop output, stands to benefit from lenient treatment by INRA officials, if he is prepared to compensate in some way for his previous misdemeanours, whilst at the same time providing accurate evidence that he is fulfilling a socio-economic function (by farming all or part of his land productively, exploiting the forest commercially or using it for ecotourism). Although a resolution of absolute nullity (given for serious violations of the law, 'the use of physical or moral violence' against the estate administrator etc.) calls for more severe handling, a landlord will not necessarily lose his land through reversion if he can prove the estate is 'productive'; instead, he will be permitted to regain possession of the property after paying a sum of money fixed by the Superintendency. This implies, as a number of experts have pointed out, that virtually any speculator can make amends for his acquisition of property by unscrupulous devices merely by ensuring that a part of the estate is exploited commercially.

Saneamiento is the third component of the land distribution rationalization programme. Clearly, the types of problems to be remedied, some of them dating from the 1950s, vary considerably. For example, properties frequently lack precise boundaries: in some cases the same lands have been 'granted' on a number of occasions, giving rise to bitter feuding over multiple ownership claims and resulting in loss of productivity. Irregularities in land titling paper-work abound and an unknown number of today's landholders occupy land registered in the names of deceased parents or other relatives. Whilst it is anticipated that it will take up to ten years to remove land distribution irregularities in all parts of the country, some areas, presenting fewer difficulties, have already been declared *sano* ('cleansed'). Unlike the owners of medium-sized properties and estates, *campesinos* and indigenous peoples are not required to pay for *saneamiento* services.

Articles 63 to 72 describe three types of *saneamiento*. Straightforward *saneamiento* is the model applied when there are no known conflicts over land ownership. It is envisaged that in some localities the procedures adopted will be more elaborate e.g. property will be surveyed and statistical data, collected. Where *saneamiento* occurs in indigenous settlement areas and original communities, the participation of the local people in discussions and decision making is 'guaranteed' by INRA and will be actively encouraged. During the actual process of *saneamiento* only the indigenous peoples and the communities can petition for land grants; once the work has been completed large estate owners and others are entitled to make offers for unclaimed land. According to article 43, *dotación* (granting land without payment to *campesinos* and indigenous peoples) 'will have preference over *adjudicación*' (adjudication, usually leading to the sale of land to large landholders and others, for a price fixed by the Superintendency).

It is highly probable that the leaders of the indigenous peoples and their confederation representatives would have marched to La Paz in August/September 1996 even had the *campesinos* decided not to protest publicly. Whilst Bolivia's Forestry Law, passed by Congress a month before the demonstration, had given 'indigenous peoples the exclusive right to use forest resources on communal land in territory reserved for indigenous peoples', it had at the same time authorized the government 'to lease forests to private logging companies in 40-year concessions providing regulations governing sustainable forest management are followed' (*LP*, 1998). Moreover, the titling promised after the 1990 march had not occurred and the August *Ley INRA* draft proposal failed to make any meaningful reference to territorial lands.

Consequently the terms of the agreement reached with government ministers on 20 September 1996 are incorporated in the final clauses of *Ley INRA*. Solón, writing in early 1997, remarked that the indigenous peoples gained more from *Ley INRA* than they had done throughout the previous 40 years of agrarian reform. The new law ratified the titles to the five territories discussed in 1990; thereafter all such lands would be referred to as indigenous community lands. Four more territories would be titled within 60 days of the law's publication and a further 16 would be processed within a period of 10 months.

Although frequent reference is made in *Ley INRA* to the socio-economic function of medium-sized properties and large estates, mention of agricultural activities is confined to the 'Final Dispositions' section. It is stated that INRA will compile a register of 'movable property', including agricultural machinery and livestock. The sole reference to any support for peasant farming is contained within one sentence. 'The State will authorize and channel credit for development and support for owners of small properties, cooperatives and indigenous communities'.

Ley INRA in action

Few laws in Bolivia's history have aroused stronger feelings than *Ley INRA*. Antezana's scathing comment about the legislation affecting *latifundistas* is at complete variance with that of Cárdenas. According to the latter, *Ley INRA* is 'much more compact and coherent' than the 1953 Agrarian Reform Law; it gives security of ownership to all types of landholder, especially the *campesino*. It guarantees access to new land and is just in that it does not require the poorest farmers to pay a land tax. Likewise, Isabel Lavadenz, the first INRA director, summarized the first year's achievements of *Ley INRA* as follows: whilst 2,800,000 hectares of land had been titled in favour of seven indigenous groups in the departments of La Paz, Santa Cruz and Cochabamba, INRA was also in the process of titling 62 *campesino* communities in La Paz, Oruro and Potosí. To what extent have negative views in general on *Ley INRA* changed since the dramatic events of 1996?

Solón's comments on the gains of indigenous peoples reflect the *Ley INRA* promises made in good faith but, sadly, not the actual reality of the situation two years later. By August 1998 a number of designated territories, including five Guaraní regions in the Chaco, were still being titled, whilst in others *saneamiento* procedures had not even begun. The representatives of 92 Chiquitano communities, in the process of 'recovering' their lands through *saneamiento* and titling, were on the verge of marching to their Monte Verde lands to physically eject 36 *hacendados* occupying more than 1 million hectares of land illegally. In April the departmental INRA director had 'sent orders' for seven 'owners' without any form of documentation to leave the area and 'notifications' to 30 absentee landlords with illegal papers. According to the prefecture, the removal orders could not be implemented because of the lack of personnel 'to exert pressure' on illegal occupants. Indigenous union leaders blamed the delays and inefficiencies largely on the lack of financial resources and excessive bureaucracy. The indigenous peoples' feelings of being betrayed yet again by the government are apparent from the following remarks, 'For us to recover our territory is like regaining our life - and to lose it is like dying' (a Chiquitano teacher): 'the only ones that do not enjoy privileges in this country are the indigenous peoples, despite being the only ones that comply with all the laws' (a Chiquitano union leader, also quoted in *La Razon*, August 1998).

Whilst many indigenous groups live under the constant threat of intrusions by logging companies - some with concessions in neighbouring areas and others without any form of contract - at least those whose lands are now fully titled have the law on their side. Thus, early in 1998, the Chimane, Tacana and Mosetene peoples gained the full support of the local municipal government, together with that of the NGO, Veterinarians Without Borders (VSF), in their campaign to ban the Berna logging company from operating in one of the areas titled by *Ley INRA*. Unfortunately, despite the acquisition of land titles, security of tenure for forest people cannot be guaranteed by **any** law. As in neighbouring countries, the difficulties of policing tracts of forested land remain insurmountable.

According to the Superintendency, the Lake Titicaca region was one of the first areas to be regularized by *saneamiento* procedures. Because of the acute pressure of population on available land resources, and the inability of *ex-hacendados* to reassert their authority, virtually no land along the lake-shore was available for redistribution purposes. This being the case, *saneamiento* was completed in months, rather than years. Most of the lakeside *campesinos* interviewed in September 1998 were adamant that they have gained nothing whatsoever from *Ley INRA*, whilst others feel slightly more secure as a result of having land titles up-dated i.e. having plots registered in their own names.

Nothing has been done by INRA to ease the widespread problems of *minifundismo* and, in some extreme cases, what has been termed, *surcofundismo*, i.e. the cultivation of single furrows in a number of different land plots. The *campesinos* themselves are the first to admit that there is no easy solution to the problems of landlessness and fragmentation, especially as few small farmers are prepared to even discuss the possibilities of moving away from the communities of their birth to farm in an alien environment. For the *campesinos* with minute plots of land, cash cropping remains out of the question: it is difficult enough to provide for their own basic subsistence needs. No incentives to dissuade young lakeside dwellers from migrating to the city have been forthcoming.

Ley INRA contains no meaningful proposals to make available financial support, including the provision of rural credit facilities, for agricultural development, or rural development in general; it makes no reference to the restoration of agricultural extension services. No undertaking has been given in connection with securing fair, realistic prices for *campesino* market produce: instead small farmers continue to be exposed to what they consider unfair competition and the uncertainties of an unregulated market. *Campesinos* complain that, whereas the 1953 Agrarian Reform Law encouraged community syndicates to play a key role in agricultural and community development, today's syndicates and cooperatives have been marginalized by *Ley INRA;* whilst it advocates the distribution of available land to communities, rather than individuals, it has done nothing to strengthen them. All lakeside communities are disillusioned by the ex-vice president's failure to fulfil his declared ambition of building a rural university, with an agronomy department, near the community of his birth.

Whilst gaining nothing, lakeside *campesinos* agree that they have lost nothing as a result of *saneamiento*. On the other hand, their fears originally aroused by *Ley INRA* have increased with the passage of time, and especially since the appointment of an INRA director with the interests of large landholders at heart. They still fear the possibility of a land tax being imposed on small farmers at a future date: likewise, the threat of land being seized for non-payment of debts is a lasting cause for concern. All *campesinos* regard the Minister of Sustainable Development and the Environment as a public enemy and, living in an area where ecotourism is actively promoted, dread the permitted expropriation of land for environmental or 'public utility' reasons. Although they were originally assured by the government that *Ley INRA* would give them greater security and full ownership, rather than working, rights to their land, they maintain that it is difficult, if not impossible, to feel secure when they can be deprived of land at any moment. Some lakeside dwellers remain convinced that they will be forced into selling land to speculators at some point, by their own poverty, whilst others are relieved that the new law has provided them with the option of selling family-owned land if they decide to join their children in La Paz or El Alto. They do, however, resent the fact that the price of land for sale is fixed by the Superintendency; they argue that they would be able to obtain higher prices if left to negotiate without interference.

In more isolated areas of the republic, *saneamiento* procedures are proving much more complicated and tedious than those already undertaken within the Lake Titicaca region. Since the publication of *Ley INRA* the Roman Catholic Church hierarchy in Sucre, together with the INRA departmental director, have brought to the attention of the public the plight of landless *campesinos* in Chuquisaca and the difficulties of regularizing the landholding situation in a number of provinces. For example, the Chaco provinces of Hernando Siles and Luis Calvo were completely 'untouched' by the 1953 Agrarian Reform Law and have never been surveyed. At the point in 1997 when *saneamiento* processing (assisted by a Dutch donation of US$ 10 million) began in the department, no institution had more than the vaguest notion of the number of *campesinos* with or without land, nor of the number of *latifundistas*, in these two provinces. According to INRA whilst recent discussions with *campesino* leaders and aerial photography have confirmed that many *campesino* families are landless, and that much of the land is held irregularly or abandoned, it will take until at least the year 2000 to complete the *saneamiento* procedures. *Campesino* leaders maintain that *latifundistas* within the provinces are continuing to exploit and maltreat *ex-colonos*, whilst 'laughing' at agrarian reform: INRA is adamant that abandoned and semi-abandoned estates will be subject to reversion and land subsequently granted to peasant communities.

Campesinos and their unions throughout Bolivia are greatly concerned, with every justification, about the power of the Agrarian Reform Commission to make decisions, which could have serious, detrimental implications for individual small farmers and *campesino* communities. In the early version of *Ley INRA*, it had been proposed that six, of the possible eight, votes would be required to amend the law but, according to the published text, only five are needed. This implies that the three *campesino* and indigenous peoples' representatives are powerless to prevent the law being modified in favour of large estate owners.

There is little doubt that INRA departmental directors will in due course dissolve a substantial number of *latifundio* holdings in isolated areas of poor quality soils (as in the Chuquisaca provinces referred to above), where there is very little incentive for absentee landlords to pay tax on unproductive land. Similarly, measures are being taken in several departments to resolve the problems arising from high-profile 'land scandals'. Bolibras, which forced the closure of the

CNRA in 1992 (p.77), again made headline news in 1997, when an INRA inspection revealed that the lands in question, supposedly returned to state ownership in 1995, had been occupied since that date by an MNR ex-senator, Oswaldo Monasterio, and his associates, including several Brazilians and a Canadian. Early in September 1998, Monasterio and his cronies were served by the Santa Cruz director of INRA with a writ requiring them to abandon all Bolibras lands within a period of 10 days.

With these exceptions, despite their original forebodings, estate owners have lost very little by *Ley INRA*: indeed, some have already obtained titles securing ownership rights and benefited financially. Speculators and *loteadores* (individuals seizing tracts of land) in the Santa Cruz region have retained possession of land acquired through fraud or influence, after agreeing to pay the reduced land tax based on their self-assessed property value. Their only concern is that under-valuation will prove costly in the event of a decision being made by the Minister of Sustainable Development to sequestrate land for environmental or 'public utility' reasons.

Latifundistas are not experiencing problems in persuading INRA officials that parts of their properties are serving a socio-economic function, since what constitutes such a function is not clearly defined in the new law. They do not have to present statistical proof that they have farmed all or part of their land commercially by reaching a certain level of productivity in terms of output per unit area; the proportion of land to be farmed is not indicated. Likewise, exactly what is meant by fulfilling a socio-economic function with regards to ecotourism provision, remains conveniently vague. In the event of an estate being declared subject to reversion, it has usually been found possible to regain possession through the payment of a sum fixed by the Superintendency.

Individuals who wish to acquire land for speculative reasons, or for specific purposes, such as building a tourist hotel, are now able to purchase land, except from *campesino* communities, at a fixed price. Whilst the president, the Minister of Sustainable Development, the Superintendent and the INRA director are known to have the interests of business men and large landholders at heart, and the powers of rural syndicates and national *campesino* unions are at a low ebb, most estate owners feel confident that little will be done to threaten their livelihoods further. The writer finds it impossible to agree with Antezana's claim (1997, p.51) that:'*Ley INRA* has the objective of making the virtually landless *campesinos* systematically lose their lands and return to being *colonos*'. On the other hand, it is hard to disagree with his assertion that the 1996 legislation is guaranteed to reinforce and consolidate the *latifundio* and limit the property rights of *campesinos*.

Clearly, *Ley INRA* is essentially a land reform law: it makes minimal reference to agrarian reform measures. Originally the Sánchez de Lozada government had considered the possibility of introducing a separate agricultural modernization bill, aimed at boosting economic growth, reducing the country's food insecurity and attacking rural poverty. In early 1996 a working party comprising politicians, economists and agrarian experts produced a consultative document outlining 'a strategy for the transformation of agricultural productivity', containing an authoritative and realistic analysis of the state of Bolivian agriculture and proposals for change. Eight objectives were identified: to promote a technological 'leap' enabling agriculture to become 'an engine of growth'; to stimulate the growth of more and better employment opportunities in both rural and urban areas; to combat poverty, improve living conditions and increase the opportunities for human development in rural areas; to increase food security; to promote the participation of sectors of the population that had not benefited from traditional development strategies, in particular the rural population, marginalized urban dwellers and the *campesina;* to implement strategies for the sustainable management of natural resources; to compensate for the historical deficit in public investment in farming, especially in the areas of technology, rural infrastructure and social services (education and health) and to devise policies for rural areas, designed to lead to decentralized, micro-industrialization.

It was argued in the document that Bolivia's development strategies throughout the 1970s and 1980s had been 'basically anti-rural, anti-agricultural': the new strategy would have as its central objective 'the reversal of this tendency, the capitalization of rural areas and the conversion of the agricultural sector into a catalyst for growth and diversification of the economy'. The four guiding forces would be: a technological thrust in agriculture, investment in human development in rural areas, the rational management of natural resources and investment in roads and irrigation, 'to expand markets, reduce risks and increase productivity'.

Such proposals reflected in large measure the recommendations made in 'Promoting sustainable agriculture and rural development', Chapter 14 of *Agenda 21*, drafted by the United Nations prior to the 1992 Rio Earth Summit. Whilst referring to agrarian reform as one of the 'main tools' of sustainable agriculture, the document strongly advocates the practice of incorporating agricultural development and environmental protection as integral components in rural development planning - a strategy rarely pursued in the Third World and unfamiliar to many industrialized countries, including the United Kingdom. Sadly, although international funding for the agrarian reform programme was procured, the strategy was shelved. The writer was assured by Ministry of Agriculture personnel in September 1998 that the recommendations contained in the 1996 publication will form the basis for an agrarian reform 'package' to be introduced in 2000 or 2001, but such an initiative seems highly unlikely under the present government. For the moment *campesinos* are obliged to rely on NGOs and university agronomy departments for agricultural advice and support.

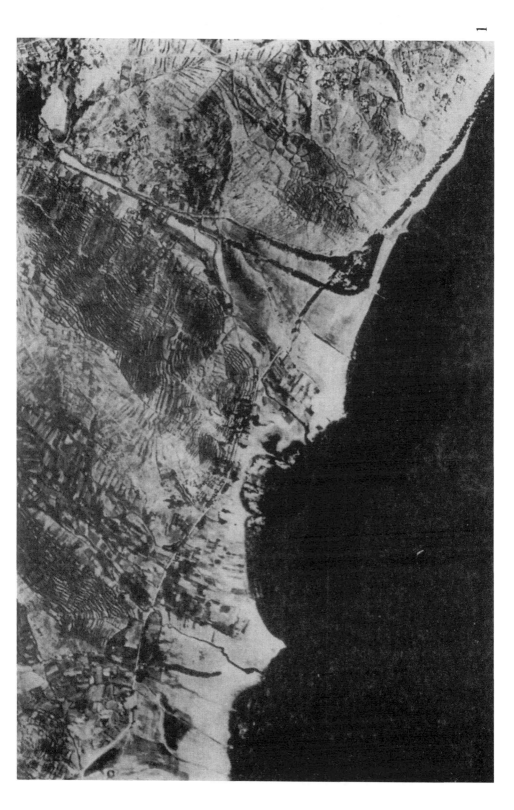

2

3

10

11

14

15

Part II:
The Practice

6 The Lake Titicaca region

When approached by the writer in 1971, SNRA research staff strongly recommended the north-eastern shores of Huiñaymarca, Lake Titicaca's smaller lake, as an ideal location for investigating the long-term impact of Bolivia's National Revolution on rural communities. Also known as El Lago Menor, El Lago Chico and El Lago Pequeño, Huiñaymarca(ka) is almost completely severed by the sharp promontories of the Santiago de Huata and Copacabana peninsulas from the main lake, El Lago Grande, sometimes referred to as El Lago Mayor or El Lago Chucuito. Access to the Peruvian part of Huiñaymarca involves crossing either the International Bridge over the River Desaguadero (marking the boundary between the two republics) or the 850 metres-wide Tiquina Straits, by means of a flat-bottomed barge or small boat - a hazardous undertaking in stormy weather.

The lakeside between the small town of Huarina (see map p.104) and the cantonal centre of Jank'o Amaya was portrayed by the SNRA as an area dominated by Aymara farming communities in which the traditional Aymara way of life and language had survived prolonged and turbulent periods of Inca and Spanish colonialism. In the last decades of the nineteenth century encroachment on Indian freeholdings and subsequent incorporation into the *hacienda* system within the area had been more extreme than elsewhere. Apart from favourable climatic, edaphic and locational factors, *hacendados* had been attracted by the existence of a dense Indian population, providing them with a ready supply of farm labour.

Since the turn of the century the lakeside region in question has been the scene of widespread Protestant, particularly Baptist, activity, suggesting that in pre-reform times not all farming communities had been isolated, self-contained units as they were widely reputed to be. In 1971 the area was one of acute

Lake Titicaca

R. Huancané
Huancané
R. Ramís
Moho
Juliaca
R. Coata
284m.
Sota
Island
Puerto
Acosta
R. Suchez
N
Escoma
BOLIVIA
Ancoraimes
Puno
Chucuito
LAKE TITICACA
Island of
the Sun
Achacachi
Huatajata
PERU
R. Ilave
Ilave
Tiquina
Huarina
Juli
Copacabana
Pomata
HUIÑAYMARCA
Peninsula Taraco
Tiwanaku
Desaguadero
Guaqui
R. Desaguadero

—10— Depth of lake (metres)

—·—·— International boundary

Barrage across the River
Desaguadero (the lake's only
outlet) constructed by the Lake
Titicaca Binational Authority

10   0   20 miles
10   0   20 kilometres

to Lake Poopó

population pressure on available land resources; yet it lay in close proximity to the country's focal point of political and economic life, to the Peruvian border and to government-sponsored colonization projects in the Yungas. It was only to be expected that between 1952 and 1971 significant modifications to traditional farming practices and living conditions, together with some measure of out-migration, should have resulted from the interactions of so many impinging forces. Not surprisingly, agricultural change rapidly emerged as the central and all-embracing theme of the doctoral research.

El Lago Sagrado

Over the centuries the Sacred Lake has attracted countless thousands of pilgrims to its islands and lakeside settlements; it continues to hold a deep fascination for travellers. According to pre-Inca mythology, the most sacred of the Andean gods, Viracocha, 'emerged from the lake in complete darkness to create the world ... man, the sun, the moon and the stars'(Gutierrez, 1997). Likewise for the Inca people Lake Titicaca represented the birth place of their civilization. Writing in the early seventeenth century, the Spanish chronicler Garcilaso de la Vega observed that:

> The Incas say that on this (Titicaca) island, the Sun placed his two children, male and female, when he sent them down to instruct the barbarous people who dwelt on the Earth...They say that after the deluge, the rays of the sun were seen on this island, and over the great lake, before they appeared in any other part (Swaney and Strauss, 1992, p.185).

Ever since its foundation the small lakeside town of Copacabana (with a population of 3,379 in 1992), on the southern shores of El Lago Grande, has drawn pilgrims from far afield. It was established by the Inca Emperor Tupac Yupanqui as a resting place on the route to the Rock of the Puma, a sacred site for human sacrifice on the island of Titicaca, later renamed Isla del Sol (Island of the Sun) by the Spanish. Since construction work began on its basilica in the early seventeenth century, Copacabana (160 km from La Paz) has also been acknowledged as an important Andean-Roman Catholic shrine. It welcomes pilgrims (some of them having walked considerable distances) convinced of the miracle-working powers of its black statue, carved in the 1580s by the grandson of Tupac Yupanqui, of the *Virgen de Candelaria*, Bolivia's patron saint. Paceños, taking their newly-acquired vehicles to receive the Virgin of the Lake's blessing and protection, fervently believe that they and their vehicles will remain accident-free for years to come.

Visitors to Bolivia and Peru can not fail to be impressed by the scenic splendour of Lake Titicaca in its striking Andean setting, by 'its serene beauty' (Osborne, 1964), by its magnitude and vivid blue waters - and, not least, by the everyday farming and fishing activities of the people living along its margins. Today's tourists from La Paz, if not making their way directly to Machu Picchu, are likely to be visiting Copacabana, having stopped briefly, or stayed overnight, in Huatajata. At Huarina, 75 km north-west of La Paz, a road continues northwards to Achacachi (with 5,602 residents in 1992), the administrative centre of Omasuyos province in which the lakeside communities between Huarina and Jank'o Amaya are situated; although Omasuyos is one of the smaller provinces of La Paz department, according to the 1992 national census its rural population of 68,101 exceeded all others. Traditionally lakeside *campesinos* made the journey to Achacachi within a day, by walking from the communities over the mountains. As already noted, Achacachi 'became the center of Aymara peasant organization' in the early stages of the National Revolution (Klein, 1992, p.235). The town retains its national reputation for unrest and belligerency. In 1971 rumours of 'men from Achacachi' coming to intervene in a feud over land in the community of Chua Visalaya provoked considerable fear amongst both factions. In September/October 1996 banners bearing the name of Achacachi featured prominently in the *campesino* protest march and 'occupation' of La Paz; it was *campesinos* from Achacachi who were seen on one occasion to be hurling rocks and lumps of concrete at passing vehicles.

Recent research on Lake Titicaca

It is now widely accepted that Lake Titicaca began life as a larger, deeper lake, known to geologists and geomorphologists as Lake Ballivián. This prototype lake had been formed during the middle-Pleistocene epoch, after the Sorata glacial phase, in an Altiplano depression, which had been deepened in Pliocene times i.e. at the end of the Tertiary period. At an altitude of 3,810m, Lake Titicaca qualifies as the world's highest commercially navigable lake. In its entirety it covers an area of 8,400 sq km, with Huiñaymarca accounting for 16 per cent of the total surface area; the lake's maximum length is 176 km, its maximum width, 70 km. Whilst Lake Titicaca is fed by 25 rivers, it functions virtually as an inland sea. The Desaguadero, its only outlet, flows southwards some 400 km, over an increasingly arid and saline Altiplano surface, until it enters the extremely salty and mineral-contaminated Lake Poopó, which in a normal summer dries up completely. Whereas the Desaguadero accounts for approximately 5 per cent of Lake Titicaca's total water discharge, between 90 and 95 per cent is lost by evaporation.

Whereas most of these facts have been common knowledge for generations, accurate statistical data, especially relating to the depths of the lake and former lake levels, have been in short supply until recent times. Thus in 1976 the findings of the first reliable binational (Bolivia and Peru) hydrographic survey of Lake Titicaca contradicted the earlier speculations of writers, such as Osborne (1964, p.10) and Carter (1971a, p.10), about a maximum depth of 1,500' (457m). It is now established that the lake's deepest point, off the Peruvian island of Sota, is 284m. Whilst water depths exceed 200m over extensive areas of El Lago Grande, the greater part of Huiñaymarca is, by comparison, very shallow, with depths of between 5 and 10m. Only in the Chua channel, two miles off the shoreline between Huatajata and Compi, does the water reach a depth of 40m.

During the last decade scientific and environmental research on Lake Titicaca has flourished. In 1991 the French Institute of Scientific Research for Development and Cooperation (ORSTOM), in collaboration with the Bolivian Institute of Social History (HISBOL), published a carefully researched lengthy volume, edited by Claude Dejoux and André Iltis and entitled *El Lago Titicaca: Síntesis del conocimiento limnológico actual* (Lake Titicaca: a synthesis of present knowledge about the lake's physical features). The book contains numerous authoritative research papers/chapters under seven headings: Genesis; Geomorphology and sedimentation; Palaeohydrology; Climate and hydrology; Physical and chemical characteristics of the water; Biological communities and Economic exploitation of the lake.

Since October 1991 wide-ranging scientific surveys have also been conducted in connection with a major joint Bolivian-Peruvian initiative to devise a strategy for regulating Lake Titicaca's water levels, which, according to records kept since 1912, fluctuate by as much as 6.3m. Whilst flooding problems had been discussed on a number of occasions since the signing of a preliminary convention in Lima in 1955, extreme weather events forced the pace in the 1980s. Droughts in 1980 and 1983 were followed by three years of heavy rainfall, together responsible for violent fluctuations in water levels, irreparable damage to property, transport dislocation, widespread inundation of farmland, crop failures and animal deaths - causing considerable hardship for lakeside communities and serious financial losses for both countries. Increased awareness and understanding of the *El Niño* phenomenon has heightened fears of such extreme weather events recurring in the future.

Once the floods had abated in 1987, Peru and Bolivia signed a research funding accord with the European Community, setting up *Plan Director Global Binacional* (Global Binational Master Plan) with the ultimate objective of regulating water levels and managing water resources within the TDPS system i.e. the enclosed basin of the Altiplano, containing Lake Titicaca, the River Desaguadero, Lake Poopó and the Coípasa Salt Lake. In 1991 a number of European consultancy firms were commissioned by *El Proyecto Especial Lago*

Titicaca (PELT) to form a consortium responsible for carrying out a series of detailed scientific studies; such surveys have enabled PELT to construct invaluable geological, geomorphological, hydrological, fluviomorphological, climatological and edaphic maps of the Lake Titicaca region.

In July 1997 the Binational Authority (ALT) began work on what is proving to be a controversial and costly enterprise to build a series of barrages across the Desaguadero. The main barrage's four outlets are to remain open during the rainy season to release excess water, whilst the discharge of water from the lake into the river during periods of drought will be restricted. According to the *Bolivian Times* (July 1997), 'by regulating the water levels, fauna and flora that depend on the lake should be better preserved as the lake will not be allowed to go above or below the levels at which the system functions best'. Clearly the barrages **could** have a significant bearing on the livelihoods of local farming communities.

The physical background of the research area: limiting factors

Geology and soils

In pre-Columbian and Spanish colonial times, invading forces, groups of settlers and itinerants made their way along the more easily negotiated south-western shores of Lake Titicaca, purposely avoiding the problematic, irregular margins of the area under consideration. Whereas the Peruvian shorelands gradually merge into an extensive, monotonous, dry plateau, interrupted only by the occasional deeply incised river valley, the Bolivian communities along the eastern fringes of the lake are hemmed in by land rising abruptly and spectacularly to the lofty summits of the Andean Cordillera Real. Permanently snow-covered Illampu (Ancohuma), at 6,429 m above sea level one of Bolivia's highest peaks, lies less than 65 km north-east of the communities. It forms an integral part of a *serrania* (the most northerly of five such ridges), of Palaeozoic massifs, trending south-eastwards from the Peruvian border to the Pass of Luribay in the environs of La Paz. Averaging some 5,540 m for more than 160 km, this ridge functions as an important climatic divide and no less as a human one, in that it presents an almost insurmountable barrier to communication and thereby isolates the eastern mountain slopes, deeply scored by tributaries draining towards Amazonia, from the densely peopled northern Altiplano and lakeside region.

The oldest surface rocks within the region date from the middle-Devonian period; this particular rock series, known to Bolivian geologists as *Serie Sicasica*, covers wide expanses of the Santiago de Huata peninsula. Mainly comprising medium-grained sandstones and quartzites, with occasional intercalating clay bands, the middle-Devonian strata are generally resistant to weathering and give rise to well defined 'hog backs', which approach the lake-shore almost at right angles. From the aerial photograph it is clearly apparent that in pre-Columbian times the Devonian hillsides were intensively and carefully cultivated: walled terraces up to approximately 4,000m obliterate the natural surface relief.

East of such exposures, partially consolidated sediments up to 30m thick are believed to represent terraces of Lake Ballivián. It is thought that towards the end of the Tertiary period, beds of mud and clay were laid down horizontally: to the residual material has been applied the local name, *Formación Chua* . During the final phase of Andean orogenic activity in Plio-Pleistocene times, vulcanicity occurred on a large scale. Fine-grained basaltic and andesitic extrusions obscure extensive areas of the Devonian strata and sometimes reach a depth of 200m. According to an SNRA survey (1954), this was a region 'of totally stony lands, rocks and scarps, unsuitable for any type of farming activity'.

For the most part, the lake margins and valleys in the region, largely corresponding to areas under cultivation, are superficially covered by Quaternary deposits. Such deposits are of varied origins: riverine, glacial, fluvio-glacial and lacustrine. In some cases rivers have spread fans of gravel around the bases of hill slopes. Elsewhere glaciers gouged out the lava plateaux and subsequently glaciers and rivers transported morainic material towards the lake. Homogeneous clays were left behind by the retreating Lake Ballivián. Inundation and associated lacustrine deposition remains an important factor in the economic livelihood of the lakeside communities. Whilst flooding by lake waters may signify temporary disaster to certain *campesinos,* the film of alluvium remaining after the water's withdrawal implies a partial replenishment of soil fertility and increased production for the following year - a pattern which is likely to be disrupted once the Desaguadero barrages are in full operation.

In a number of communities east of Chua considerable accumulations of gravel and large stones make cultivation impracticable. In the past large stones were sometimes used as field markers or in the construction of animal enclosures: of recent years, large stones and gravel from the Huatajata area have been ruthlessly exploited for road building purposes, creating a variety of problems discussed later. Clay soils within the region have over the centuries provided the basic material for *adobe* houses, outhouses and walls; unless irrigated in the dry season, such soils become concrete-like and seamed with deep cracks. Elsewhere the soil takes on a loamy appearance and, in certain localities, sand (from the erosion of sandstones and quartzites) has been deposited by streams and wind to form the characteristic soil. Although the alluvial clays between the lake and road are generally regarded as the most productive soils, crop cultivation on land under the direct influence of the lake has always involved an element of risk; at certain times of the year it is likely to be waterlogged, whilst during the dry season it is impregnated with salt. In common with the soils of the northern Altiplano, those within the study area are generally deficient in organic matter.

103

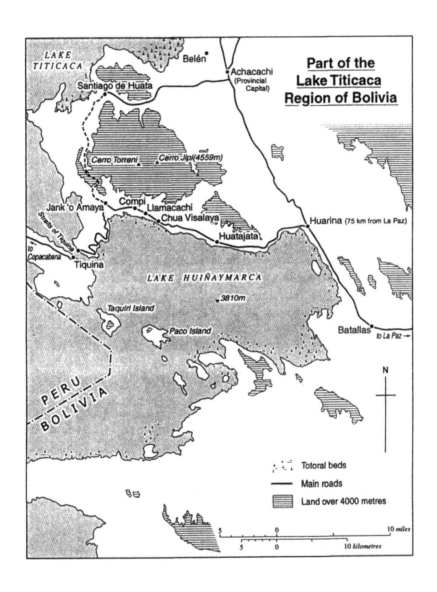

Part of the Lake Titicaca Region of Bolivia

LAKE TITICACA

Belén

Santiago de Huata

Achacachi (Provincial Capital)

Cerro Torreni

Cerro Jipi (4559m)

Jank'o Amaya

Compi

Llamacachi

Chua Visalaya

Huatajata

Huarina (75 km from La Paz)

Straits of Tiquina

to Copacabana

Tiquina

LAKE HUIÑAYMARCA

3810m

Taquiri Island

Paco Island

Batallas

to La Paz →

PERU

BOLIVIA

N

Totoral beds

Main roads

Land over 4000 metres

5   0   10 miles

5   0   10 kilometres

The 1954 SNRA survey of Chua classified the region's soils as generally 'first-grade'. Ten years later the Agricultural Bank's survey assessed the lakeside soils as a mixture of 'third-grade' and 'fourth-grade', noting that: 'From the point of view of its chemical composition, the lack of analysis makes it impossible to classify the type of predominant soil'. It is abundantly clear that since the early 1970s the overall quality of the soils within the area has deteriorated markedly. Today's problems of severe - if not critical - land degradation are discussed in the last chapter.

Relief and drainage

Along the northern margins of the Santiago de Huata peninsula, where the growth of *totora* (lake reed) is well developed, an already expansive plateau surface is being gradually extended by lacustrine sedimentation. Conversely, along the southern rim of the peninsula the terrain rises so suddenly and steeply from the lake that at certain points, such as Chua Cayacota, it has been difficult to find sufficient space between the hillsides and the land liable to flood for constructing even a narrow road. With the exception of the immediate shores and several valleys, such as those of Compi and Chua Visalaya, widening as they approach the lake, level surfaces are a rarity. Where they do occur, as north of Mt.Jipi (4,559m), they are virtually inaccessible, frequently waterlogged and well above the normal limits of cultivation for most temperate crops. Within the region as a whole the general impression is one of extreme ruggedness. Around the steepened slopes of mountains such as Jipi, where erosion and frost shattering are potent forces, there are large rock-strewn patches and in places sheer vertical cliffs of andesite are exposed.

The area has a dendritic drainage pattern, with all its rivers and tributaries gravitating towards Lake Titicaca, which acts as the local base level of erosion. The majority of watercourses are intermittent i.e. they only flow during the rainy season and when invigorated by the melting snows of the *cordillera*. More often than not, streams marked as 'permanent' on topographical sheets hardly merit the classification. For example, the one draining Chua Visalaya is extremely sluggish during the late winter months, and in years when the summer rains are delayed degenerates into a series of disconnected pools; on the other hand, in times of exceptionally heavy precipitation and sudden rainstorms, it can overflow its banks and even transport eucalyptus trunks.

Until the 1980s a lack of potable water (water for domestic purposes being obtained from open earthen wells) and very poor standards of hygiene contributed to abysmal health standards throughout the lake region. Dysentery and typhoid were commonplace. Not surprisingly, several of the women interviewed in 1971 stated that their basic problem in child rearing arose from the necessity of having to wash young children in cold, contaminated water. At the same time MACA attributed occurrences of *distomatosis* (riverfluke) along the lake margins to the existence of stagnant green water in disused or badly maintained ditches, in river beds and hollows, usually created by making *adobe* bricks. Although since the early 1980s (starting with Chua Cocani in 1981), CARE International, in collaboration with CORDEPAZ (the former departmental development corporation), has installed potable water supplies in virtually all lakeside communities, water for agricultural usage remains a bone of contention.

In former times disagreements over the allocation of water from rivers and streams, together with the diversion of water courses, were commonplace. Feuds between individual families and between different communities not infrequently led to bloodshed. Water contamination from the washing of clothes upstream was often a source of annoyance to farmers nearer the lake. More significant was the blocking of streams during the dehydrating of potatoes and *oca*, severely limiting the quantity of water available for irrigation purposes downstream. The situation has been exacerbated since the early 1990s by the introduction of readily available *bombas de agua* (motorized water pumps). Whilst a limited number of relatively wealthy *campesino* families, with the aid of *bombas* and lengthy plastic tubes, now find it an easy task to regularly irrigate their crops and in so doing drain accessible water courses, the mass of the poorer *campesinos* are obliged to physically carry water considerable distances to their fields.

Limitations imposed by climate

In its 1995 strategic plan ALT applies C.W. Thornthwaite's climate classification to areas within the TDPS system. The lakeside region is indicated as experiencing a C(o,i,p)C' climate i.e. a sub-humid climatic type (associated with grassland vegetation) with a microthermal temperature efficiency and a prolonged virtually dry period (late autumn/winter/early spring). Average annual precipitation is said to vary between 600mm and 800mm, whilst an average temperature of 9°C is experienced - a meaningless statistic without qualification.

Clearly, as elsewhere on the Altiplano, temperatures are considerably reduced by altitudinal effects. Whereas the Brazilian Mato Grosso town of Cuiabá (16°S, 56°W and 166m above sea level) has an average monthly temperature of 27°C, the corresponding figure for Copacabana (16°S. 69°W, but 3,843 m above sea level) is only 9.4° C, despite the benign influence of the lake. Additionally, nearness to the north-west to south-east aligned Andes is a decisive factor in terms of precipitation: the Santiago de Huata peninsula lies directly in the rain shadow of the Cordillera Real and is consequently deprived of rain from the south-east trade winds.

The relationship between Lake Titicaca and weather patterns in the immediate vicinity is complex. By reason of its vast size, the lake assumes the nature and properties of an ocean or sea. With a constant temperature of 10°C, it exerts a moderating influence, enabling crops to be grown at far higher altitudes than would normally be practicable; it also stimulates plant growth by increasing precipitation. Whilst minimum temperatures for January (i.e. mid-summer) are almost identical for lakeside Copacabana and Belén, several miles from the lake, winter temperatures contrast markedly: Copacabana has a mean July temperature of 8°C and a mean minimum of 2.8°C for the same month, whilst the comparable figures for Belén are 3.6°C and -5.9°C. Away from the direct influence of the lake the actual temperature range widens and the dangers of frost damage to crops increase significantly. Even within a short distance of the lake, diurnal temperature range, especially in the dry season, is considerable: in the day time there is intensive insolation from a cloudless, luminous sky, whilst at night rapid radiation occurs and cold air drains down to valley bottoms, which act as frost pockets.

Rainfall is strongly concentrated in the months from October to April, normally reaching a maximum in the months of December, January and February; 83 per cent of Huatajata's rainfall occurs between October and April, 56 per cent of it in the three months from the beginning of December to the end of February. Total rainfall decreases as the lake's influence diminishes: thus, whilst Huatajata has a mean annual rainfall of 789 mm and Copacabana, 788 mm, the corresponding figure for Belén is only 477 mm. Whilst rainfall is higher in the whereabouts of the lake, the highly seasonal and relatively unreliable nature of precipitation constitutes a major problem in the absence of comprehensive irrigation systems. Drought remains a major hazard to lakeside cultivators and even short delays in the summer rains can imply serious hardship in the following year. According to the ALT document: 'The losses from droughts in the past decade are more than five times greater than those produced by lake flooding, in spite of the extraordinary character of the latter' (which it was acknowledged had destroyed 3,367 houses and seriously damaged 4,135 others in La Paz department alone). The manner in which precipitation occurs is also a significant factor: sudden rainstorms cause flashing, temporary waterlogging on hillside terraces, gully erosion on other hillsides and much rain is rapidly dissipated through thin, absorbent soils.

Though Lake Titicaca provides more favourable conditions for agricultural activities than those encountered elsewhere on the Bolivian Altiplano, even apart from flooding, it can indirectly wreak havoc in lakeside communities. The lake creates its own system of land and sea breezes; strong outflowing day-time winds disturb the lake surface, sometimes making fishing hazardous and prohibiting crossings at the Straits of Tiquina. In the early evening, land breezes, predominantly northerly and north-westerly in direction, frequently approach gale force. In October 1971 the writer witnessed such a freak wind (referred to in national newspapers as a *huracán*) develop over Huatajata; it killed numerous animals, destroyed crops and razed more than 40 *adobe* buildings to the ground. If damage on this scale can be caused by local winds, unprotected thin and fragile soils are highly vulnerable, as are cereal crops valued for their above-ground growth. *Tormentas de graniza* (hailstorms) constitute a hazard of similar magnitude and are also more prevalent in lakeside localities than further inland. Copacabana experiences an average of six hailstorms per year, usually between October and April. In the lakeside communities hail is dreaded and considered as a curse on the community for which the spirit world must be appeased. In one example observed, the only 'proof of whether or not a girl had induced an abortion, thereby committing an act 'against the laws of nature', was held to lie in whether or not there occurred a damaging hailstorm in the ensuing month. It is significant that more *brujos'* (sorcerer) supplications are concerned with drought and hail than with any other community problem. At the same time, being struck by lightning is one of the conditions enabling an individual to train as a *yatiri* (Aymara word for sorcerer).

Thus, whereas the lakeside communities enjoy a more congenial climate than their counterparts on the dry, bleak, windswept Altiplano, the natural base of life is none the less precarious: social, political and economic conditions may change but the *campesino* remains under the constant threat of natural hazards, especially lacustrine flooding, drought, hail and strong winds. Whilst measures to improve irrigation could diminish fears of drought, hail and 'hurricanes' are uncontrollable phenomena and to date no serious attempt has been made by individual communities, or groups of communities, to prevent the damaging effects of flooding, since longer-term soil enrichment is believed to compensate for temporary discomforts.

Natural vegetation and cultivated crops

More favourable climatic conditions than those encountered elsewhere on the Bolivian Altiplano are largely responsible for the relatively rich natural vegetation of the lakeside region. In contrast to the isolated clumps of coarse, spiky *ichu* grass (*paja brava*), characteristic of so much of the *puna*, *xerophytic* flora along the lake margin is abundant and varied.

Whilst herbs and natural grasses predominate, two native shrubs thrive in lakeside communities. *Quishuara* (sometimes referred to as 'the wild olive of the Incas'), requiring very little water and able to withstand strong, biting winds, flourishes on hill slopes. It is tended as a garden plant, along with the bush bearing the colourful *kantuta*, Bolivia's national flower. Additionally, *yareta*, a shrub preferring volcanic soils, and *thola*, more suited to sandstones, grow on the Devonian and basaltic-andesitic outcrops. On the lower hill terraces and especially on the sheltered sides of stone-walled hill tracks, herbs grow in profusion. All have local Aymara names and many have been highly prized for generations for their medicinal properties: some are widely respected as curatives for muscular pains and heart disorders, others are used in bathing young children, one is fed to dry cows to induce lactation, and at least three herbs are imbibed in *maté* form to terminate unwanted pregnancies.

The Agricultural Bank's 1964 survey of Chua Visalaya identified five dominant native grasses, including *cebadilla*, wild barley. All five grasses are of medium coarseness, become progressively parched and brown in the dry season and are of little value as animal fodder. (The SNRA surveyors had concluded in 1954 that a cow, without any additional sources of forage, would require at least seven hectares of pasture land for grazing purposes, and each sheep, at least one hectare). Traditionally, dried grass was used for thatching purposes and as a fuel for cooking, but in an age of zinc roofs and kerosene stoves, it is valued for binding together *adobe* bricks, for storing dehydrated potatoes and *oca*, for packing eggs for market and for protecting nursery beds of onions from inclement weather.

Totora beds, fringing the lake and varying considerably in width, retain their value for *campesinos*, despite the fact that some of their traditional uses e.g. for making reed boats, thatching houses and mattresses, have all but disappeared. In pre-agrarian reform times local *hacendados* exercised complete control over the gathering of *totora* growing between the boundaries of cultivable *hacienda* land and the lake. In the freeholding Indian communities, such as Llamacachi (whose *totora* beds reach a maximum width of almost 80m), patches of reeds were and still are, owned by individual families; less fortunate *campesinos* are obliged to buy *totora* by the donkey-load. When green, *totora* stalks, measuring up to 3.5m in length, can be cut for animal fodder. *Campesinos* insist that, although not as nutritious as *alfalfa*, *totora* is more beneficial than most natural grasses in the region. In former times, during periods of drought and enforced food shortage, lakeside families were sometimes reduced to eating the pith of *totora* stalks. Decaying stalks and stubble are burnt to provide a fertilizing ash and - according to lakeside dwellers - induce renewed growth. An additional important source of cattle and sheep fodder (especially for *campesinos* lacking direct access to *totora* beds) is a protein rich aquatic macrophyte, *chancco* or *llachu*, which grows on the lake floor. Like *totora*, *chancco* thrives in slightly saline water: it is removed from the shallow waters of the lake and trailed to the shore, with the aid of a boat and long poles.

All of the native food plants previously referred to - *papa* (*Solanum tuberosum*), *ulluku/papalisa* (*Ullucus tuberosus*), *oca/oqa* (*Oxalis tuberosa*), *quinua/kiuña* (*Chenopodium quinoa*), *cañahua/qanawi* (*Chenopodium pallidicaule*) - in addition to *isañu* (*Tropaelum tuberosum*) and *tarwi/tarhui* (*Lupinus mutabilis*), are believed to have been grown in the lakeside region in pre-Columbian times; they are still cultivated today, though the relative popularity of crops has changed. Throughout the region the 'sweet' potato remains the unchallenged dominant staple crop. To indigenous varieties have been added a number of improved types, such as *Sani Imilla*: whereas the latter are marketed, local varieties are generally consumed by the producers because of their preferred taste. Although some of the more bitter potatoes (such as *papa amarga/papa luki*), tolerant of frost and alkalinity, were traditionally cultivated on stone-faced terraces, most potatoes today are grown in lakeside fields and in adjacent valleys, especially near buildings where frost is considered less of a threat. Whilst some potatoes are consumed fresh and crops of improved varieties are sold in local or city markets, large quantities of traditionally grown potatoes are processed, preserved and eaten in dehydrated form as *chuño* (black in appearance) or *tunta*, (white). *Oca*, second only to the potato in the *campesino* diet, and also eaten in dehydrated form (*c'aya*), is often grown on land unploughed after the harvesting of a potato crop. *Isañu*, a tuber rich in almedin, is grown under similar conditions to *oca* but less than in former times.

The grain crops *quinua* and *cañahua* yield abundantly in the lakeside region; apart from being able to grow at an altitude of 4,000m, they are well adapted to withstand both frost and drought conditions. Although in past centuries *cañiwaco* (*cañahua* flour) was used for making bread, cakes and biscuits, it is mainly used today to enrich soups. *Quinua*, believed to have been domesticated 5,000 years ago and the mainstay of the Incas' diet, is currently attracting considerable attention in North America and Europe, because of its remarkable nutritional value (p.33): during recent years disputes over Canadian and US attempts to patent hybrid strains have been extremely bitter and remain unresolved. *Quinua* 'surpasses other cereals and grains with its essential amino-acid composition. In other words, this ancestral American food possesses proteins of superb quality' ... it contains 'high quantities of vitamins B, C and E' in addition to various minerals, and '*quinua's* high calcium content make it an ideal milk substitute' i.e. once its saponin has been removed (Vargas, 1997). *Quinua* is widely used by lakeside *campesino* families in soups: additionally, old *quinua* stalks are burnt and the resultant ash is added to coca to make the stimulant more appetizing and harden the tongue. Like *quinua*, *tarwi*, a lupin cultivated in the Andean region at least 1,500 years ago and traditionally consumed in soups, is at present the subject of international interest and scientific investigation. The plant produces a grain with a 50 per cent protein content and is also a potential source of oil but, so far, it has been found difficult to counteract its element of toxicity. Of recent years it has been incorporated in cereals: thus, one 'instant multi-cereal', available in markets in La Paz is advertized as containing wheat, soya, *amaranth* (a highly nutritious pre-Columbian grain grown in the more southerly Tarija area) and *tarwi*. *Tarwi* is also sold as a cure for diabetes, rheumatism and neuritis.

Of the non-indigenous food plants introduced to the Lake Titicaca region during or since Spanish colonial times, barley and hybrid varieties of beans and onions have fared best. Whereas in the early 1950s wheat was being sold commercially (to the brewery in La Paz) by the *hacendado* of Chua and was also being grown for bread by Canadian Baptist missionaries in Huatajata, today's *campesino* production is insignificant; as previously noted, wheat is only grown with difficulty at heights above 3,500 metres. By contrast, barley is still regarded within the region as an important cereal, although it is cultivated primarily for animal fodder and straw. Sometimes barley and beans are grown in alternate rows, the latter providing protection from damaging strong winds; in other situations, barley is planted around nursery onion beds, to protect the young plants whilst producing extra forage. Improved varieties of beans, such as *Hacienda* and *Francesa*, were introduced to the lakeside in pre-agrarian reform times: they are valued as food for local consumption, as a cash crop and their stubble is burnt to return nitrogen to the soil or used to supplement fodder supplies.

Although Andean varieties of onions have been cultivated in parts of the Altiplano since pre-Columbian times, the possibility of marketing them as a lucrative cash crop was not recognized in the lakeside area until after agrarian reform, once the seeds of improved strains became available for purchase in local markets. Near the lake onions take only four or five months to reach maturity and can be planted at any time of the year i.e. the optimum season for planting can be carefully calculated so that harvesting coincides with known shortage periods in city markets. Whilst onions provide the main source of income for numerous family units in the study area, their cultivation is tedious; it involves a considerable amount of hand labour and depends on the regular and careful application of water. In lakeside communities which are not heavily dependent on marketing onions as a cash crop, crop rotations have changed very little since pre-Revolution times. Traditionally a four year rotation system was strictly observed by households in *comunidades indígenas* and *colono* families cultivating *sayañas* on *haciendas* . Crops of potatoes (*papa*), were followed by *oca*, *oca* by beans (*haba*), and either barley (*cebada*) or wheat (*trigo*) completed the sequence. *Hacendados* followed a modified rotation system, replacing *oca*, the native crop, with barley or oats (*avena*). Although *hacienda* owners produced crops continuously in the fertile soils bordering the lake, on poorer quality land at some distance the four year course would be followed by a fallow period lasting two or three years - a luxury denied to *colonos* farming stony hillside fields.

Many families in the study area continue to follow the same basic pattern, except for abandoning the cultivation of wheat and integrating crops of *quinua*, *cañahua* and *tarwi* or *isañu*. In Llamacachi and Compi onions have not displaced the four major crops but have extended the rotation period by two years in order to accommodate the cash crop: (1) potatoes, (2) *oca*, (3) onions, (4) beans, (5) barley and (6) onions: not infrequently two crops of onions are produced during the year on the same patch of land.

111

According to Dejoux et al. (1991), the most popular four year rotation practised around the shores of Lake Titicaca today is as follows: (1) potatoes, (2) beans, (3) 'cereal or *quinua*' and (4) 'a forage cereal, generally barley'. The ALT document (1995) refers to the dominant sequence within the TDPS as a whole as: (1) potatoes, (2) *quinua*, (3) barley and (4) *tarwi* or beans. Xavier Albó notes that rotation systems differ according to region, climate and access to water supplies: despite such variations, the major feature i.e. the cultivation of potatoes in the first year, never changes - it has remained the same over centuries. He also refers to the widespread practice of inter-cropping to protect weaker plants from frosts, to control the spread of diseases and to use nutrients more effectively. The most common plant associations he identifies as: 'beans-barley, beans-*quinua*, barley-peas, barley-*quinua*, *oca-isañu* and *oca-papalisa*' (1989, p.32).

7 The state of agriculture in lakeside communities on the eve of agrarian reform

The north-eastern shores of Huiñaymarca had acted as a powerful magnet to late nineteenth century expansionist landlords. Excessive abuses of the *hacienda* system since that time, together with the lakeside dwellers' traditional reputation for belligerency and the communities' nearness to La Paz, made it inevitable that the study area should become the scene of social unrest and violence during the early days of the National Revolution. It was a region that experienced the full impact of agrarian reform. Not surprisingly, the majority of household heads interviewed in 1971 were able to recall aspects of their working lives as *colonos* vividly .

Within the general region the communities of Chua Visalaya and Llamacachi (between Compi and Huatajata) were suggested by SNRA personnel as meriting in-depth study of their response to the widened range of economic opportunities opened up by the implementation of agrarian reform. Chua Visalaya and Llamacachi are contiguous lakeside communities, displaying similar physical features: their hillside terraces provide dramatic visual evidence of pre-Columbian settlement and cultivation. The communities are bisected by the main La Paz to Copacabana road, which marks the boundary of the *orillas* (the lake-shores), highly valued for farming purposes but at the same time vulnerable to lake flooding. Whilst the 1950 national census had not referred specifically to either community, their populations in 1971 were comparable. Chua Visalaya contained 72 households and a total resident population of 326 (166 males and 160 females); additionally, 37 Visalayans (19 males and 18 females) were living outside the community, mainly in La Paz. In Llamacachi 278 people (132 males and 146 females) occupied 63 houses; all but a few of 92 migrant Llamacacheños (55 males and 37 females) had also taken up permanent residence in the primate city.

113

Despite their shared characteristics and the passage of time, the *campesinos* of Llamacachi and Chua Visalaya remained strongly conditioned in a variety of ways by their contrasting histories. In pre-reform times Chua Visalaya had been an *estancia* (section) of the larger Chua *hacienda*, whilst Llamacachi even to this day prides itself on being one of the very few communities in the lakeside region to have survived intact as a *comunidad originaria*.

Including a freeholding in the analysis of agrarian reform made it possible to broaden the comparisons of agricultural productivity and economic efficiency in pre-reform times: what has normally been referred to as a landlord and peasant 'dual economy' was, in reality, a 'triple economy'. Whereas *colonos* were deliberately prevented by *hacendados* (if not by lack of time and access to cultivable land) from engaging in marketing activities, *comunarios* were free to travel, to diversify their production, to follow marketing pursuits and to seek waged labour opportunities beyond the confines of their community; likewise they were at liberty to join political organizations and their access to education was not impeded. It was interesting to probe the extent to which they had exploited such opportunities in pre-reform days. Whilst the majority of *comunarios* did not benefit directly from land expropriation and redistribution, had they made use in the years following agrarian reform of the new opportunities offered by expanding marketing facilities, the availability of improved seeds, chemical fertilizers etc. and colonization projects? Had the *comunarios'* sense of determination, stubborn perseverance and initiative - qualities engendered over centuries of struggle and resistance - enabled them to maintain an economic advantage over their previously down-trodden, demoralized neighbours, with little, if any, experience of the outside world, or had the economic differences between them rapidly faded with the implementation of agrarian reform?

Whereas community boundaries beyond the immediate influence of the lake were in *hacienda* times - and still are in many cases - a bone of contention and often ill-defined, property boundaries along the cultivated *orillas* were generally very distinct, as clear from the aerial photograph of the district, taken in 1956 i.e. three years before the process of *dotación* was completed in Chua. The photograph illustrates the marked difference in land use between the *hacienda* and the *comunidad originaria*. Llamacachi's myriad of small, cultivated lakeside fields are flanked by the extensive single-crop *hacienda* fields of Chua Visalaya on the eastern side and Compi to the west. Settlement patterns were also strikingly different in *hacienda* times. Whilst Llamacachi's 'ribbon development' along the road had become well established, in Chua Visalaya the *hacienda's* dual economy and class divide were reflected in the *colono adobe* housing units, clinging precariously to the hillsides (above the *hacienda's* valley pasture lands) and the ostentatious *casa hacienda* (*hacienda* house), built on elevated land above the road, which was flanked by eucalyptus trees planted in the 1930s.

Contested space

The freeholding of Llamacachi is enclosed by the much more extensive *ex-haciendas* of Chua and Compi and the lake. The community's history of resistance against successive *hacendados* of both properties was recounted by Hans and Judith-Maria Buechler in *The Bolivian Aymara* (1971), based on their anthropological survey of Compi. In the mid-nineteenth century Llamacachi had been one of four sections of the much larger community of Tauca, at a time when Compi proper and Capilaya were functioning as separate *haciendas*. Towards the end of the century the two *haciendas* merged and launched a series of sustained attacks against the *comunarios*, in order to gain control of *comunidad* land. 'Only Llamacachi and a handful of Tauca and Kalamayo peasants succeeded in maintaining their independence'. The Buechlers (1971, p.5) refer to an incident occurring at the turn of the century - a story which has been handed down from one generation to the next:

> One night, the *patrón* [the *colonos'* term for the *hacendado*] was about to invade the house of Manuel Ramos. Hearing of this, Manuel quickly put on women's garb and thus escaped undetected to Achacachi, the province capital. There he asked for aid, returning to Tauca with both soldiers and a judge. The *patrón* was warned against any further abuses and the judge then appointed Manuel as *jilakata* [the highest official] of the community which had become leaderless. From that day on, no more sales of land to Compi occurred.

Llamacachi also had to fend off the encroachment advances of its other neighbour, Chua. At one stage a particularly ruthless *hacendado* exercised such compulsion that all but one family sold their land to the *hacienda*. However, over a period of some nine years, 'the Llamacacheños created so much trouble for their landlord that he finally sold most of their land back to them'. In later decades landlords of both Chua and Compi tried to intervene from time to time in Llamacachi's internal affairs, especially to forestall political activity and prevent the construction of a school - both of which could have acted to their disadvantage. In Llamacachi cooperation was not forthcoming. As one *comunario* pointed out to the writer in 1971, 'we refused to bow down to the landlords like donkeys'. Whilst Llamacachi is to be admired for its audacious spirit and fortitude throughout the long period of expansion and consolidation of *haciendas*, the internal feuding over so many years was to leave an indelible mark. As the Buechlers remarked in 1971, land disputes 'are one of the roots of the traditional mistrust with which *comunarios* and *colonos* view each other'.

So far as is known, no serious attempt had been made to establish even the total dimensions of Llamacachi in pre-Revolution times: according to Carter (1964, p.5), a detailed plan of the community of Irpa Chico (south-west of La

Paz), prepared for him by an SNRA surveyor, was the first to have been made of 'a free community on the Altiplano'. Likewise a general lack of demographic data made it impossible to establish the man-land ratio in Llamacachi in pre-reform days. Whereas Burke claimed (1970, p.421)that since the National Revolution the population of the lakeside *ex-haciendas* studied by him had increased by between 50 and 100 per cent, it seemed in 1971 that the reverse had happened in Llamacachi i.e. that the process of rural-urban migration had already begun to make its mark. According to the *comunarios* interviewed, several families and individuals had moved permanently to La Paz in the 1950s and 1960s, because of land shortages and disputes over land ownership.

Whilst it was frequently claimed that the pressure on land resources had been almost as intense as in neighbouring *haciendas*, freeholdings such as Llamacachi had obviously enjoyed certain advantages in the decades preceding agrarian reform. *Comunarios* had not had to live under the constant threat of being deprived of land, whereas on some estates, *aynoka* plots (community lands cultivated individually but in strict accordance with a prescribed rotation system) could be confiscated for 'misdemeanours'. (In Chua Visalaya, one *ex-colono* claimed that he had been deprived of usufruct rights after the deaths of several diseased sheep during his term as *camana* (guardian) and a widow, recently returned to the community, alleged that her family had been forced to leave the *hacienda* because her husband had attended the First National Indian Congress in 1945). Furthermore, Llamacacheño families, unlike lakeside *colonos*, had been able to farm the highly-valued tracts of flat land adjacent to the sacred lake. Some families claimed to be in possession of land titles granted as a result of the 1854 Indian reviews but this could not be substantiated: because of countless inter-family land disputes (involving accusations of stealing, removing stone boundary markers etc.) no householder was prepared to produce written evidence of ownership. Members of the two largest landowning families alleged that even before the National Revolution, holdings had been considerably fragmented by repeated subdivision of *parcelas* in accordance with inheritance rights. No *comunario* could recall a time when there had been an annual or any other regular form of land redistribution; by tradition, plots had remained in the same families and rarely changed hands.

Although Llamacachi's area of cultivable land only comprises some 80 hectares, in actual fact the community's tract of intensively cultivated, highly productive land south of the through-road exceeds that of its neighbour, Chua Visalaya. The 80 hectares referred to by no means represent the total area devoted to arable farming by *comunarios*; since pre-reform times *comunarios* have been devising strategies to obtain access to additional plots. For example, the father of the writer's field assistant was in 1971 cultivating two plots of land in Llamacachi itself, two in Compi, two more in Cawaya (another section of Compi) and one in Chua Visalaya - all as a result of mutually agreed *anticrético* arrangements, whereby creditors enjoy usufruct rights to certain plots of land (belonging to the debtors) in partial payment of financial debts. Likewise the greater part of uncultivable land in Chua Visalaya was in 1971 being almost as widely grazed by Llamacacheño livestock as by animals raised by *ex-colonos*; *comunarios* also traditionally had access to the pasture lands of neighbouring Compi. Clearly, whether referring to environmental or economic factors, it is meaningless to approach communities in the lakeside area - as elsewhere on the Altiplano - as units of land separated from each other by rigid boundary lines.

Although it is likely that much of the lakeside area was granted in *encomienda* (protectorate) in the early days of Spanish occupation and made the transition to *hacienda*-type ownership in the seventeenth century, the earliest document preserved in the files of Chua's *expediente* (the dossier compiled by the SNRA as a prerequisite for land redistribution) refers to the year 1832. At that time Chua was owned by a certain José Ballivián, assumed to be General José Ballivián, under whose leadership 'the Bolivians experienced the greatest success they were to enjoy in their entire history' i.e. the defeat of the Peruvian forces invading the department of La Paz in 1841 (Carter, 1971a, p.41). Significantly, Ballivián was one of a handful of large landowners before the twentieth century to appreciate the nutritional value and economic potential of indigenous crops. As Carter observes, 'he pushed the development of *quinua* as a cash crop and established a *quinua* bank to provide loans to growers of that cereal'.

From the *expediente*, it appears that all seven Chuan *estancias* remained under single ownership until 1939, when the property was divided. Whilst unsuccessful attempts had been made in the late nineteenth century to acquire the coveted lake-shore land of Llamacachi, elsewhere the *hacienda* met with better fortunes. Thus in 1957 two adjoining communities (Pacharía and Pallareti) forwarded petitions to the SNRA claiming that between 1900 and 1952 certain of their lands had been usurped and forcibly incorporated into *Hacienda* Chua by successive *hacendados*. Likewise Corpachilaya made a claim against the estate and was in 1959 awarded 46.6500 hectares of land previously expropriated.

Chua's *expediente* refers indiscriminately to the *hacienda*, *propiedad* and *estancia*: it also contains conflicting statistical data, especially with reference to the size of the estate. The figures occurring with the greatest frequency indicate that Chua Visalaya/Chua Cocani (the entire property in the 1950s) comprised 5599.7650 hectares; Chua Visalaya contained 2276.5425 hectares of land. On the property plan of Chua Visalaya, drawn by an SNRA surveyor in 1957 (see p.136) the land was divided as follows:

Area cultivated by the *hacienda*	40.8700 has
Area cultivated by *colonos*	87.4423 has
Area of *aynokas*	630.0000 has
Area occupied by school and sports field	2.5950 has

Area of uncultivable land   1515.6352 has
                                                          Total   2276.5425 has

By way of comparison, the four Bolivian lakeside *ex-haciendas* studied by Burke in the mid-1960s averaged 3670 hectares. According to him, the Bolivian *haciendas* 'were, to a much greater degree than the Peruvian ones, mere agglomerations of small Indian *sayañas*' (Burke, 1971, p. 314). The statistical data produced by the SNRA led Carter (1964, p.71) to conclude that: 'Ownership of a *hacienda* in pre-reform days could not be taken at its face value ...Though the landlord had legal title to its entire area, he was in fact limited to an impressively small portion of it. The rest was exploited by and for the peasants'. Whereas the *sayañas* of Chua Visalayan *colonos* occupied 87.4423 hectares, only 40.8700 hectares - less than two per cent of the total area - were cultivated by the *hacendado*.

Bare facts of this nature can easily create misleading impressions. Acute population pressure on available land resources and excessive fragmentation of *sayañas* were endemic characteristics of lakeside *haciendas*. According to the 1957 plan of Chua Visalaya, of the 87.4423 hectares classified as *sayañas* or *terreno de ex-colonos* (land of *ex-colonos*), 42.1063 were held by 110 families in Visalaya, 22.7600 by 34 Carapatans and 22.5760 by 71 Cayacotans i.e. the dwellings, outhouses and land parcels of as many as 215 families were limited to a similar number of acres. In addition, 34 *ex-colono* families were classified as landless. (Since only 72 families were living in Visalaya in 1971, it seems likely that the 1957 SNRA statistics did not refer exclusively to household heads).

Whilst the *colonos' aynokas* occupied 630 hectares of land, large sections of this zone lay above the limits of cultivation for even the hardiest of temperate crops, were covered by thin stony soils and, in any case, presented almost insuperable irrigation problems. A 1956 SNRA document, concerned with the parcelization of land commented as follows: 'All of the *aynoka* lands... are excessively subdivided because *colonos* and their sons work the same plots. In this zone the parcels cultivated by each of the *campesinos* have become extremely fragmented with the result that some *campesinos* hold larger plots and others, very reduced ones'.

In view of the various discrepancies in statistical data it is clearly impossible to deduce precise man-land ratios in pre-reform days; at the same time the excessive pressure of population on cultivable land is indisputable. *Colono* plots were indeed highly fragmented: one Visalayan had 16 *kallpas* (small plots), another, 15 - and cultivation of these scattered patches was marginal. Some *campesinos* said in 1971 that one reason why, as *colonos*, they had wanted large families had been in order to be able to satisfy more *hacienda* obligations thereby gaining access to additional plots in the *aynoka* lands.

118

Unqualified statistics disguise the fact that in Chua Visalaya, as on every other *hacienda*, the *hacendado* invariably cultivated the most fertile and potentially productive sections of the estate. One could be forgiven for assuming that an *hacienda* on which only two per cent of the land was cropped by the landowner must unquestionably have fallen into the category of the 'backward', non-mechanized, unproductive *latifundio*. In the case of Chua Visalaya, nothing could have been further from the truth. The last *hacendado* of Chua, Gordon Barbour, farmed the estate so progressively that it became one of the most efficiently operated enterprises in Omasuyos. It was inevitable that *Hacienda* Chua should eventually be declared an *empresa agrícola*, which it was in 1959.

Apart from winning national championships with his livestock and mechanizing the estate to an almost incredible degree by Bolivian standards, Barbour had conducted crop experiments in collaboration with agronomists from the nearby agricultural research station of Belén. In 1950 he rented the property of Compi; subsequently he mechanized the estate and reduced the hours of labour required of the *colono* families. He it was who inaugurated today's flourishing yachting club in Huatajata and on the lake shores of Cayacota constructed a pier and established a well-equipped boatyard, by 1950 employing several skilled mechanics from La Paz and 30 local men, a number of them from Llamacachi. Despite their antagonistic feelings towards local *hacendados*, Llamacacheños admitted to having respected the last *patrón*: one *comunario* remarked that 'the *patrón* worked his men hard but he was a fair-minded man and Chua was the best-run estate in Omasuyos'.

Whilst Chua Visalaya's *colonos* had many features in common with *colonos* of the majority of *haciendas* in the lakeside region, clearly the *hacienda* emerges as atypical in terms of its management and agricultural productivity. It offers an unusual opportunity to reflect on what lakeside properties were - and are - capable of achieving under imaginative but sound management, with the benefit of agricultural research, with adequate capital investment and the cooperation of a trained work force.

The dual economy of pre-agrarian reform Chua

(i) *Hacienda Chua*

Whilst in the pre-reform period there were marked differences between the economic opportunities and living conditions of Llamacacheño *comunarios* and Chua Visalayan *colonos*, contrasts **inside** Chua's *hacienda* were far more striking. No lakeside community could have exhibited a more clearly defined dual economy than did Chua in the 1940s and early 1950s. Much of the descriptive and statistical data used by way of illustration does in fact date from 1954, during which year a comprehensive SNRA survey of the estate was conducted. Five more years were to pass before the process of *dotación* was completed: some *campesinos* later alleged that legal proceedings were delayed by the *patrón's* bribing of the secretary general of the newly created syndicate. The estate's only immediate response to the Agrarian Reform Law Decree was the introduction of remuneration in the form of cash; after August 1953, the *hacendado* paid each peasant worker the stipulated 350 *bolivianos* per month. In all other respects the *hacienda* functioned much as before until 1959.

The fact that Supreme Resolution Number 84494 declared the *hacienda* to be an '*empresa agrícola ganadera*' (an agricultural livestock enterprise), 'consolidating in favour of the property an area of 3101.1700 hectares', suggests that stock rearing activities had been predominant on *Hacienda* Chua. Certainly, considerable importance was attached to animal husbandry and large sections of the valley, retained exclusively for the grazing of the landlord's livestock: such a preference for pasturing animals in a central place, where soils could be carefully protected from infection and contact with diseased creole stock avoided, was more than understandable in view of the high quality of the *hacienda's* cattle and sheep. On the other hand, the 1954 SNRA report stated that arable farming provided the main source of income. *Expediente* documents clearly indicate that both arable farming and stock rearing were well organized, modernized, highly productive and lucrative activities during the *hacienda's* final years.

Unfortunately, only very rarely were reliable records e.g. on crop output per unit area, kept on Bolivian *haciendas*. It is impossible to draw comparisons between landlord and peasant productivity with any degree of accuracy. Whilst the Chuan *expediente* recorded the 1954 crop production figures for the two *estancias* no reference was made to the actual area under cultivation and the quantities themselves pose insoluble problems. Total crop output was given in quintals as: 3000 qq. of potatoes, 8000 qq. of barley grain, 1500 qq. of broad beans and 1000 qq. of oats. Assuming one quintal to be the equivalent of 100 lbs., the following tonnages can be derived: 134 tons of potatoes, 357 of barley grain, 67 of broad beans and 45 of oats. Dion in 1950 assessed the average potato yield per hectare on the Altiplano to be in the order of 2.5 tons. Burke, writing 20 years later (1970, p.418), computed average productivity statistics for the major crops in the lakeside communities of his survey. It is useful to refer to his example of potato yields for the purpose of comparison. He deduced that Bolivian *campesinos* were at that time obtaining on average 4,494 lbs. of potatoes from each hectare planted: the corresponding figures for Peruvian lakeside *campesinos* and *hacendados* were given as 4,758 and 10,073 respectively i.e. even the highest figures represented little in excess of 4 tons per hectare. Relating these figures to Chua's 134 tons of potatoes, one is bound to conclude that: (a) a much larger area was in fact under cultivation than portrayed on plans of the *hacienda* (which seems unlikely), (b) yields in Chua were exceptionally high or (c) the figures are wholly inaccurate. The truth probably lies in modifications of (b) and (c): Barbour may have used 1q to represent 10 lbs or 10kg. Barley statistics are even more suspect, since it was widely agreed amongst *campesinos* that the main emphasis in the *hacienda* had always been on potatoes: a 1948 telegram concerns a shipment of 40 sacks of seed potatoes from New York (via Mollendo, Peru). Wheat was also grown on the property; a bill also dated 1948 specifies 460 kg. of Klein Success grain and the same quantity of *Sinvalacho*, sent from La Paz.

Most *hacienda* produce was sold directly from the *hacienda's aljeria* (storehouse/shop) in La Paz. In former years it had been transported by mules and llamas - a journey taking trusted *colonos* at least two days to complete - but in the years preceding the National Revolution crops were delivered by the *patrón's* Chevrolet truck; in the 1950s a second vehicle was acquired for the express purpose of transporting *hacienda* crops, dairy produce and meat.

From whatever angle the *hacienda* is viewed, it is clearly apparent that, at the time when agrarian reform measures were being debated in La Paz, and the means of raising agricultural productivity discussed, *Hacienda* Chua was being farmed intensively, on extremely progressive lines. A glance at the 1954 inventory of estate machinery suggests an amazing collection of mechanical equipment - at a time when some *hacendados* did not even own a tractor. Additionally Barbour was conducting plant experiments e.g. in 1954 he planted 10.5 hectares of *alfalfa* to supplement supplies of animal fodder - but concluded that the plant was ill-suited to local climatic conditions. Attention was also being paid to the need for replenishing soil minerals. Chemical fertilizers and pesticides were being lavishly applied: the *expediente* contains various bills for large quantities of superphosphates from the Bolivian Chemical Institute. To a certain extent, specialization of labour was in operation; two *ex-colonos* had been trained as tractor drivers. Irrigation channels were extended and preserved in a good state of repair; one 3 km-long channel, with transverse ditches, adequately irrigated most *hacienda* land under cultivation in Chua Visalaya.

Highly successful livestock production was partly attributable to the initial purchase of prime pedigree animals e.g. in 1945 the *patrón* had imported 25 Romney Marsh and 5 Oxford Downs sheep from Argentina. Efficient animal husbandry on the estate was equally significant. Such a situation was also exceptional. According to Burke (1970, p.427), 'unlike the present-day Peruvian *hacendados*, the pre-reform Bolivian *hacendados* owned a much smaller proportion of the number of livestock on the estates. In addition their sheep and cattle were nearly all of the degenerate *criollo* breed'. The Bolivian Ministry of Agriculture's 1946 survey of 39 *haciendas* in Omasuyos, Inquisivi and Los Andes, found that, with the exception of two *haciendas* 'all the remainder had only 800 Merino and Corriedale sheep and virtually no better breed of cattle'.

Sheep assumed prime importance in the *hacienda's*, pastoral activities. According to the animal census of November 1954 Chua Visalaya/Chua Cocani possessed in all 2270 sheep; none of these were less than 50 per cent pedigree and 201 were of first class pedigree Romney Marsh stock. The flock was systematically split into seven groups; two of these were normally pastured in Visalaya and the other five in Cocani. *Campesinos* interviewed in 1971 frequently remarked that 'the valley was full of sheep'. A rigorous selection was made annually to separate for sale any animals with defects; all sheep were dipped twice annually and a 'hospital' permanently maintained. Wool was sold regularly to *mestizos* from Huatajata and Huatajatans also visited the estate to purchase carcasses.

At the time of the SNRA census, 70 pedigree cattle (43 Holsteins and 27 Brown Swiss) were owned by the *hacendado*. Of the total number, three were bulls (including the national Holstein champion) and 46 were dairy cattle i.e. the emphasis was on rearing cows for milk and other dairy produce, rather than bullocks for beef or ploughing purposes - in marked contrast to Aymara practices. Pigs were also of prime quality: 5 Duroc Jersey boars had been procured from Cochabamba and the *hacienda* also numbered 72 pedigree sows amongst its livestock. Large consignments of anti-swine fever vaccines were obtained from New York in 1950 and 1954. According to the report, all animals were being fed a supplementary ration of bran, grains, sunflower cake and hay.

In combination with its pastoral activities, the hacienda operated a small dairy adjacent to the *hacienda* house. One estate inventory refers to '6 cheese presses with weights': *ex-colonos* claimed that in the early 1950s 10 large 'Argentinian' cheeses were made daily from sheep milk - this accords with the *expediente's* statement that about 300 cheeses were being sold on a monthly basis. Butter was also made in the dairy by the *colonos'* wives. Dairy products were subsequently sold to middlemen from Huatajata or taken by jeep to the *aljería* in La Paz.

Such heavy investment in livestock, good seeds and tubers, fertilizers, pesticides and medical treatment etc. was rare in pre-reform times. It was difficult to ascertain how far modernization was due to the activities of one *patrón*, a man of dynamic personality (in addition to considerable wealth) and the extent to which management had been efficient in former days: unfortunately, the *expediente* contains few details of *hacienda* operations before 1944, when the last *patrón* had gained control (through marriage) of the estate. It is unlikely that he was motivated by an eagerness to improve the hard lot of the *indios* in his care. Indeed some *campesinos* alleged that the *patrón* had been eager to finance local *fiestas* because excessive drinking led to brawls and it was in his interest to promote internal discord - provided it did not interfere with *hacienda* work - diverting attention from the demands of *hacienda* labour. Llamacacheños were convinced that Chua Visalaya's internal and seemingly insoluble friction, culminating in a violent confrontation between rival cooperative and syndicate in August 1971, was attributable to the *patrón's* direct provocation of antagonism between family groups in the last years of the *hacienda*. On the other hand, like many other *hacendados*, he had realized that to be assured of the *colonos'* optimum cooperation it was essential to avoid stretching the *colonato* system to its uttermost limits. Unquestionably, his *colonos* benefited from some of his investments e.g. the provision of animal medicines to prevent the spread of certain contagious diseases through creole stock.

The last *patrón* appears to have displayed far more interest in the actual organization and day-to-day administration of the estate than was typical. It has been estimated that on the Altiplano '90 per cent of the large landowners were absentee landlords who lived in urban centers or abroad and left operation of their farms to managers' (Clark, 1970, p.4). By the twentieth century there were few, if any, lakeside *hacendados* without stately mansions in La Paz or other Bolivian cities and many were involved in lucrative businesses or professions, only visiting their estates at harvest time. This they did in order to ascertain the quantity of produce due to them and thus be able to hold their *mayordomos* and *colonos* accountable for its transference to the *aljerías* in La Paz. The only lakeside example encountered of landlords having lived permanently on their estate was the special case of Huatajata, an *hacienda* which had been purchased (together with its 47 *colono* families) at the turn of the century by Canadian Baptist missionaries; by 1942 obligations to the estate had been discontinued and soon after titles of holdings had been transferred to individual peasant families.

(ii) *The subsistence agriculture of the colonos*

The *colonato* system was basically an arrangement whereby an *hacendado* granted usufruct rights to small plots of terrain and permitted his tenants to pasture livestock on waste tracts and certain fallow lands; in return, the *colonos* and their families were compelled to work without remuneration for the estate, supplying animals and tools when necessary, and were also required to fulfil various personal and domestic obligations for the landlord and his family, both in the *casa hacienda* and the city house.

In *Hacienda* Chua, as was usual in *haciendas* of the Bolivian lakeside region, *colonos* belonged to one of four categories depending on the number of obligations they fulfilled and the amount of weekly labour their families provided. A few Visalayan *campesinos* said that they had been *mayorunis* (working one day per week and consequently granted access to minute parcels of cultivable land) and others, *yanapacos* (with access to larger plots in return for performing additional tasks). By far the majority of household heads had been either *personas* or *media personas*; frequently, as his family grew and its labour potential increased, a *colono* had graduated from *media* to full *persona*. Chuan *ex-colonos* stated that in order to retain access to nine hectares of land, four members of a *persona's* family had as a rule been expected to contribute three days non-salaried work per week (every Monday, Tuesday, Wednesday and sometimes Thursday) and to provide full-time labour at peak periods, such as planting, weeding and harvesting: for the *media persona*, who was entitled to cultivate 4.5 hectares, the duties had been halved. *Ex-colonos* claimed that on *hacienda* work days they had found it necessary to rise before 4 am to complete their own farming chores before walking to the *hacienda* fields.

Nine hectares of land might appear to have been sufficient for the subsistence needs of a peasant family, but certainly in the case of Chua the figure should not be taken at face value; no *colono* had access to an amount of cultivable land even approaching 9 hectares. The 1957 SNRA plan previously referred to provides ample evidence of the critical pressure on land resources available for *colono* exploitation. Apart from marriage to a Llamacachi *comunaria*, one of the few means of acquiring land was to assume part of the work load of another *colono* in exchange for the usage of a plot to which the 'employing' *colono* had access. But in most cases, even had the *colono* been able to gain access to other parcels of land, it would have been of little advantage: *hacienda* tasks engaged most of his working time and he often found it extremely difficult to complete farming operations in his own fields, especially since peak agricultural periods on the *hacienda* naturally coincided with those on his own holding.

In addition to spending at least three days of each week working for the *hacienda*, every household was compelled to take its turn at major rotating tasks. Some of the particularly time-consuming duties were as follows: *camana* (guardian) of *chuño* and *tunta*; supervisor of *totora* beds; *awatiri* or shepherd - in Chua an individual family was responsible for one group of sheep for the entire year, during which period members of the family lived in a large *adobe 'casa de ovejas'* (sheep house); *mitani* (cooking and cleaning in the *casa hacienda*); dairy duties involving making butter and cheese; *apiris* transported produce from the estate to the city; *pongueaje* included tasks ranging from messenger service to construction and maintenance work on *hacienda* buildings. Failure to fulfil *hacienda* obligations successfully (e.g. by loss of goods *en route* to La Paz or wastage in the preparation of *chuño*) entailed fines of varying degrees of severity. On occasions a *colono* was made to pay a forfeit but the normal practice was for animals or produce to be expropriated. In Chua Visalaya *campesinos* insisted that punishment was repeatedly of a physical nature; there were numerous references (sometimes supported by Llamacacheños) to the usage of a silver whip by the last *mayordomo*.

On the *sayañas*, *colonos* were theoretically free to grow whatever crops they liked but this was, hardly a liberty since they were forced by circumstance to produce what was considered most essential to family needs and could not afford to observe the traditional fallow period. *Papa, papalisa, oca, quinua, isañu, cañahua* and broad beans were grown with most frequency. The contrast between farming methods on *hacienda* lands and on *colono* plots could not have been more extreme. On the *sayañas* the traditional wooden foot plough (*taclla*) was normally used: the more fortunate *colonos* used oxen for ploughing (sometimes these were borrowed by neighbours who agreed to supply a day's fodder). Otherwise a primitive digging-stick and a clod breaker (a stone attached to a stick handle) were the only implements in general usage; in any case, field size and the uneven nature of the terrain dictated hand labour. Old and unimproved varieties of seeds and tubers were planted, regardless of whether they were already diseased. Crop production on the terraced hillsides was entirely dependent on the summer rains; if they failed or rainfall was inadequate, the result could be disastrous in terms of immediate food shortages and the lack of seeds for the next year's planting. There was no possibility of purchasing fertilizers, even had they desired to do so; farmyard manure was of greater value as a fuel and used in cooking. The only natural fertilizer added to the soil was sheep *taquia*: this was collected in *bayeta* (coarse cloth) sacks from the higher pasture lands. Bean stalks had to be burnt as fuel or fed to livestock; lakeside *totorales* belonged to the *patrón* and passage to and from the totora beds across *hacienda* fields was prohibited. Whether or not the *colonos* appreciated the need for conserving soil fertility, it was impossible to do so under the prevailing arrangements; hence soil degradation became inevitable. One can only hazard a guess at crop yields; according to *ex-colonos* they were considerably less than those of the *hacienda*.

Unfortunately no pre-reform inventory was made of livestock owned by *colono* families in Chua. *Ex-colonos* insisted in 1971 that they kept more animals than in *hacienda* days; even so the 1971 total number of sheep did not begin to approach the number formerly kept by the *hacendado* alone. Apart from a few sheep (lambs were occasionally given to *colonos* in return for outstanding services as shepherds), all animals kept by the tenant population had been of inferior quality. They were grazed on the areas of common pasture in the *aynokas* and in zones marked '*incultivable*' on various hacienda plans. The only animals apparently kept in greater numbers in *hacienda* times than in 1971 were llamas; informants said that 'wealthy' *personas* had sometimes owned more than ten llamas, these being used primarily for transporting produce from the fields to the *hacienda* warehouse (a building adjacent to the *casa hacienda*). No *campesino* stated that, as a *colono*, he had been in possession of a *burro* (donkey) - as had frequently been the case in Llamacachi. Animal losses from disease were very numerous but accepted as inevitable. Women asked in 1971 whether or not they had in *hacienda* times woven (shawls) for exchange or sale for cash outside the community, answered that, due to a shortage of sheep and llama wool and insufficient time, they had been unable to weave enough *bayeta* to clothe even their own immediate families adequately.

During the 1940s seasonal migration to the Yungas had become the customary way of life for many Huatajatan men; peasants would leave the community to work for minimal wages on *haciendas* where labour was in short supply, thereby temporarily alleviating pressure on land and food in their home community.

Huatajata's unusual historical background clearly favoured this freedom of movement. Similarly in Compi, temporary migration had been a well established feature of *hacienda* life, largely because the landlords of Compi had owned property outside the region. Thus, during their term as *apiris*, Compi's *colonos* were sometimes sent to the *patrón's* other estate near Sorata, north of Achacachi; additionally, groups of lakeside *colonos* went to Sorata from time to time for prolonged work periods, whilst even by the 1930s, a number 'were always on the road' travelling with mules and transporting sheep carcasses, llama meat, potatoes and *quinua* from Compi to the Yungas (Buechler, 1972). Other Compeños migrated to La Paz because of the land shortages combined with the regime of severe repression imposed by the last owner of Compi. After 1953, the return of these *émigrés*, for the purpose of claiming land they had at one time utilized, was to foment bitter disputes. Thus in relation to Huatajata and Compi, the claim made by Buechler (1972, p.309) that, 'Analysis reveals a number of continuities from the past, which were the foundations upon which present-day regularities were built. Most important of these were the linkages between different ecological zones' - is irrefutable.

However, in Chua pre-reform migration appears to have played a minor role; in fact with the exception of the case previously quoted (the result of an eviction rather than a spontaneous movement), *ex-colonos* were unable to recall examples of pre-reform migration. This situation could have been due partly to a lack of initiative bred of excessive paternalism (considered later); but the fact is also significant that, although many *ex-colonos* insist they were 'worked almost to death', conditions were undoubtedly less severe than in pre-1953 Compi.

Many Chuan *ex-colonos* openly admitted to night-time robbery of *chuño*, broad beans etc. from *hacienda* fields, in the full awareness that discovery could result in loss of access to land and even to eviction: according to them such produce rightly belonged to them and sometimes offered the only means of averting starvation. Whereas the *hacienda* was primarily market-oriented, Chua's *colonos* never found it possible to think in terms of other than subsistence requirements. Indeed, no *campesino* interviewed had in *hacienda* times regularly sold produce in local markets; such transactions had been restricted to isolated exchanges (sometimes at *fiestas*) of small quantities of eggs, cheese and meat. To all intents and purposes, the *colono* was 'an economic hermit'; he had virtually no contact with the world outside Chua except on the rare occasions (normally twice each year) when it was his turn to convey *hacienda* produce to the city.

(iii) *Agricultural productivity in the freeholding*

In many respects cultivation methods in pre-Revolution Llamacachi resembled those practised by the *colonos* of neighbouring *haciendas*. The same crops were grown on the *comunarios'* house plots and lower hillside terraces as on the

126

colonos' sayañas, whilst the normal four-year cropping sequence was observed in fields bordering the lake, as was the case on the *colonos'* patches of *aynoka* land. Specific dates were assigned for farm activities traditionally associated with ritual ceremony; the threshing of *quinua* occurred on the day of Corpus Christi, whilst all cultivation and weeding ceased at Candlemas (2 February), since any further activity was deemed 'detrimental' to growth (Buechlers, 1971, p.11). Likewise potatoes were regarded as the mainstay of the diet and economy; *campesinos* interviewed in the community in 1971 asserted that in former times 'to be rich meant to have many fields and big potato crops'.

In common with lakeside *colonos,* Llamacacheños planted unimproved seeds and tubers, storing them until needed; potatoes had been small and often blighted. Most of the *campesinos* were of the opinion that crop yields, whilst well below those of *Hacienda* Chua, had exceeded those reached on *colono* plots of land. This was only to be expected since, although an equilibrium between man and land was impossible in view of the dense population and acute pressure on cultivable areas, Llamacacheños were growing crops on higher quality soils and had access to larger supplies of natural fertilizers; most *comunarios* fully appreciated the direct relationship between the latter and crop yields. In addition to burning stubble from their *totora* beds, some families had applied sizeable quantities of sheep dung to their fields; this was collected from hill pastures by *colonos,* in exchange for small amounts of potatoes or *oca.*

Adequate supplies of water for irrigation purposes during dry periods was - and still is - a contentious issue throughout the region. In the case of Llamacachi, *comunarios* had no means of controlling the quantity of water extracted from the stream as it flowed through the *hacienda* before reaching the freeholding's terrain. On the other hand, the poorest families whose only fields were adjacent to the lake, endured considerable hardship (especially if they had been unable to preserve any of the previous year's harvest) when land was inundated as a result of the lake flooding. On such occasions they would attempt to supplement their food supplies by working on one of Compi's *estancias* when casual labour was required e.g. for the threshing of beans or barley in June (between the days of St. John and St. Peter).

Comunarios had not owned pedigree stock in pre-reform times. By the 1950s they no longer raised llamas, which were considered within the community as 'poor men's animals' - because they were unable to carry heavy loads : produce was usually transported by donkeys or mules. It had never been necessary for Llamacachi to keep large numbers of oxen, whereas Chuan families had been obliged to provide the *hacendado* with animals for ploughing purposes before the estate was mechanized. A number of *comunarios* owned one animal and borrowed another from a neighbour for team-ploughing. Some *campesinos* believed that more pigs had been raised in pre-reform days; they were bought by dealers from Huatajata and subsequently traded in the Yungas. Most households had kept guinea pigs for consumption at family *fiestas*. Although women made small cheeses from sheep milk, there was no tradition of making larger cheeses for marketing. Sheep's wool for weaving and knitting purposes was sometimes obtained from *colonos* in Compi, generally in exchange for sacks of potatoes, *oca* or *quinua*. Forage presented *comunarios* with fewer problems than it did Chuan *colonos*: livestock was freely pastured on Compi's hillsides and when fodder was in short supply, particularly in the winter months, animals were grazed along the waterfront, their diet being supplemented by *totora* stalks and *chancco* from the lake.

In Llamacachi, as in all lakeside communities, marriages had been negotiated with land and livestock prospects the prime parental consideration. Several informants bemoaned the fact that: '*no habia amor aqui en el campo*' (there was no love here in the countryside), implying that personal feelings assumed no importance in arranging marriage contracts. Under normal circumstances, a young couple would live with the male's family before building their own dwelling in the family compound. One or several animals would traditionally be given in dowry by the girl's parents: the couple would be allowed to cultivate a few furrows of land in one or all of the family *chacras* (fields) or would share-crop with the father and inherit land on his death, the amount depending on the number of children in the family. In some cases girls received smaller portions of land, whilst illegitimate children were usually disinherited.

Whilst Llamacachi had remained essentially subsistence-oriented, by the early 1950s many *comunarios* were engaging in a variety of marketing transactions, taking advantage of their roadside location and the freedom to travel denied their *colono* neighbours. Whenever harvests were sufficient to provide a reasonable surplus, the more fortunate families would sell potatoes, *oca* and *chuño* in Paceño markets. Between 1937 and 1953 11 major markets were established in the city, facilitating regular peasant marketing activities. In the 1930s all produce had been transported by mule but, according to *comunarios*, three trucks were making several journeys per week between Copacabana and La Paz by the early 1940s. Travel by truck became the established means of transport after *Hacienda* Compi acquired two vehicles in about 1947. Llamacacheños commented that this was one of the main reasons for women taking over from men as the principal vendors. Whereas cash was rarely used in any transactions involving local *colonos,* most marketing by Llamacacheñas in the city was on a cash basis - although bartering continued at local *fiestas* and *ferias* (the weekly fairs of Jank'o Amaya, Huarina and Achacachi) for essential items, such as coca, salt, wool, dyes and earthen pots.

Whilst contacts between the freeholding and markets in the Yungas had not been as widespread as those involving Compeños, they had developed to a certain extent in the years preceding the National Revolution. For example, in order to finance *fiestas* a few non-Baptist vendors made infrequent visits to market centres in the Yungas to sell dried meat.

Comunarios with unimpeded access to Lake Titicaca and to *totora* for making reed boats, also operated as small-scale fishermen. By the late 1940s Llamacacheños were fishing from 'marias' (named after the Virgin Mary) i.e. *totora* boats with only one pointed end. *Campesinos* alleged in 1971 that the *maria* had been unique to their community; it was as useful as a two-pointed boat, preferable for gathering and trailing *chancco* to the shore and could be made with much less effort. At a time before trout-salmon, *pejerrey* and *ispi* had been added to the lake's indigenous fish stock, Llamacacheños and Huatajatans were netting small edible fish, such as *boga* (*challwa*), *mauris, such'i* and *k'arachi*: catches were consumed in the communities or bartered at local markets. As previously mentioned, by the early 1950s some *comunarios* were working in the Chuan 'dockyard'. The foundations of the post-reform economy were undoubtedly in place.

Unquestionably, prior to the 1952 National Revolution, the status, economic productivity and poverty level of peasant families in the Lake Titicaca region, as throughout Bolivia, were determined to a large extent by the *hacienda* system. Those communities, like Llamacachi, that had managed to avoid being totally overwhelmed by the injustices and cruelties of the institution, or like Huatajata, that had been released from domination, remained essentially subsistent but, in many cases, had begun to exploit their relatively advantageous situations. In the lakeside area under consideration, access to education together with the influence of the Huatajata-based Canadian Baptist mission, accentuated the differences between *haciendas* and free communities.

By the 1950s most Huatajatans and at least half of Llamacachi's *comunarios* were members of the Baptist Church and this was having a detrimental impact on traditional Aymara structures. Whilst the *fiesta* system had been generally accepted as a means of eradicating marked differences in family assets, Baptist converts were no longer prepared to spend hard-earned cash on large quantities of alcohol and contribute their surplus foodstuffs; they preferred to make investments of a more durable nature. Refusal to participate in drinking bouts at *fiestas* and the end of elders' meetings inevitably created discord inside the community and also between Llamacachi and the *estancias* of Compi. Even in the early 1970s Llamacacheños were indiscriminately categorized as 'anti-social' *campesinos*. Some had in the interim withdrawn their church membership, because of the problems associated with drink.

In 1943, much against the wishes of Compi's *hacendado*, the Baptist mission built a primary school in Compi and shortly afterwards, another in Chua Visalaya. According to the mission's records, over 60 children's names had been entered on the original roll of Chua's school. Not surprisingly, few pupils had attended with any regularity; girls had frequently been made to stay at home to care for younger children, cook or herd animals . The last *patrón* of Compi actively obstructed the opening of an all-grade elementary school and would not permit the children from his estate to attend the one in Huatajata. According to *ex-colonos*, the *hacendado* of Chua did not encourage attendance at the Baptist school but did not forcibly prevent it. They insisted, however, that if any *colono* or *colono's* child was discovered communicating in Spanish, he was whipped by the *mayordomo* .

The importance of education, as a means of acquiring a basic knowledge of accountancy and communication skills in Spanish, to facilitate marketing activities - and more generally, to enable individuals to gain entry to the outside economic world - was fully recognized in Llamacachi. In the 1940s a few enterprising *comunarios* had even invited teachers into their homes to instruct local children. However until 1958 most Llamacacheños were obliged to rely on the primary schools in neighbouring communities. Any boy who was keen to be educated further had to undertake the daily return journey on foot to Huatajata to attend the Canadian Baptist mission secondary school, which today proudly boasts an ex-vice president as one of its former pupils (see p.78).

8 The impact of the 1953 agrarian reform legislation

General implementation problems

According to the SNRA, 7,322 of the 15,332 land claims presented after August 1953 had been fully processed by 1970; of the outstanding cases approximately 4,000 were awaiting presidential approval, 3,000 were still under SNRA review and the remaining 1,000 were in the early stages of enquiry. The procedures leading to a final settlement were often complicated and extremely tedious; strongly contested cases could continue for ten years.

An unknown number of landlords succeeded in intimidating *colonos*, threatening them with retribution, including eviction, if they made approaches to the SNRA. Elsewhere *campesinos* were bribed with the promise of higher wages or offers of share-cropping. On some of the more isolated estates, *campesino* leaders were, for various reasons, reluctant to assume responsibility for openly challenging the authority of *hacendados*. Without *campesino* syndicates i.e. the legal representatives of communities in negotiations with government personnel, presenting petitions was very difficult. Additional confusion and delays were related to the inability of claimants to read or understand legal documents.

In the early years of agrarian reform the processing of claims was severely handicapped by an almost total lack of suitably qualified and available personnel. 'Since the revolutionary government had nearly eradicated the national army in favor of peasant militias, the Army Corps of Engineers could not be used in the countryside; as a result the National Agrarian Reform Service came to oversee every phase of expropriation including its technical aspects' (Clark, 1970, p. 40). In the event, retired experts had to be recruited as trainers and members of the survey teams. Delays in titling were eased after 1968, as a result of the acquisition of an IBM computer and the commissioning of mobile brigades, each consisting of 'one judge, one secretary, seven topographers, one agronomist and one soil analyst,..with adequate transportation and supportive facilities' (Carter, 1971b, p.246).

Once in the field, survey teams faced the unenviable task of attempting to reconcile the demands of agrarian reform legislation with realistic conditions in the countryside. Countless inter-community boundary lines were bitterly disputed and ill-defined, as were the dividing lines between the regional zones referred to in the Agrarian Reform Law. Agrarian reform legislation had set maximum limits to smallholdings; rarely did this present assessors with any difficulty. On the contrary, in densely populated areas, such as the Lake Titicaca region, properties were, with few exceptions, too small to adequately satisfy the needs of all *campesinos* over the age of 18. Determining the status of large landholdings was not always as straightforward as might be expected. Landlords adopted a number of strategies to persuade SNRA personnel that their properties should be designated *empresas*. Lakeside *campesinos* were able to identify *haciendas* on which landlords had 'borrowed' pedigree livestock or mechanical equipment to prove they had modernized production and 'farmed progressively'.

Mobile teams frequently encountered antagonism. In lakeside communities surveyors were on numerous occasions obliged to abandon their tasks and withdraw for indefinite periods, rather than face the verbal abuse, physical obstruction and violent tactics (usually stone throwing) of resentful, suspicious *campesinos*. Internal and inter-community land disputes, multiple claims to individual plots of land and attempted land grabbing by returned migrants, or outsiders with no entitlement, often prolonged the process of *dotación*, as did the surveyors' long distance travel to communities on unsurfaced roads, some of them unusable in wet weather conditions.

Throughout the 1950s and 1960s the financial resources at the disposal of successive governments for implementing agrarian reform were minimal. 'The takeover of the mines drained massive sums from the state coffers' (Klein, 1992, p.236). As a result of 'the dramatic devaluation of national currency', Bolivia experienced 'one of the world's most spectacular records of inflation from 1952 to 1956'. The decline between 1952 and 1954 of 'the Central Bank's foreign exchange reserves...from $29.3 million to $11.6 million' was partly attributable to the state of agriculture in the early days of the National Revolution and agrarian reform (Thorn, 1970, p.177). Largely because of the widespread violence and unrest in the countryside (occupying much of the *campesinos'* time and energy), the breakdown of the *hacienda* system and the abandonment of properties by landlords, supplies of farm produce to city markets plummeted; consequently the government was obliged to expend limited cash resources on 'massive food imports to prevent starvation' (Klein).

Thus the financial resources available for the non-land redistribution components of the agrarian reform programme and for generally supporting peasant farming e.g. by providing credit and extension services, improving roads and expanding marketing facilities, were negligible. The situation was alleviated somewhat in 1955, when 'the Inter-American Agricultural Service initiated a supervised farm credit program for the purpose of mechanization, crop and livestock improvement and marketing' (Clark, 1970, p.21). The fact that the national government itself would have experienced great difficulties in contributing towards the compensation of ousted landlords, served to release *campesinos* from their financial obligations. On the other hand, they were expected to assume the main costs involved in *dotación* i.e. the payment of lawyers and surveyors. As Clark observed, 'these were considerable expenses in most cases, particularly when a complicated legal battle was contested by the landlord and revised many times ... If each member of each union had not contributed time, produce, and cash, there would have been little land reform to date [1970] in Bolivia'.

Land reform on the Altiplano

Inevitably the process and general impact of land redistribution varied markedly from region to region: environmental factors, locational aspects (e.g. distance from main communication lines and major marketing centres), historical backgrounds, population densities and levels of political awareness accounted in large measure for significant differences both between and within geographical regions. The most comprehensive survey of post-reform land redistribution on the Altiplano was conducted by Clark in 1966. The 51 *ex-haciendas* of his study, scattered between Oruro in the south and Lake Titicaca in the north, ranged in size from 335 to 9,408 hectares and contained an estimated 5,400 families in 1953. In pre-reform days all but five of the properties had been in the possession of absentee landlords, all with property in La Paz, where many of them ran highly lucrative private businesses or worked professionally. Whilst 44 of the total number of *hacendados* together held 159 estates, very few *colono* families owned even small plots of land. In the months leading up to the signing of the Agrarian Reform Decree - a period of intense political unrest and organized peasant mobilization throughout the region - virtually all the landlords and estate administrators had retired to their city residences. According to Clark, on eight of the estates in question the redistribution (by peasant syndicates) of *hacienda* land amongst *colonos* was already a *fait accompli*.

Subsequently, 25 of the 51 properties were expropriated entirely, as unproductive *latifundios*. 'The other 26 were declared medium or small properties, with the landlord retaining some land for his use. Thus the peasants on 25 farms ... realized a substantial increase in access to land while increased access on others would have ranged from substantial to nothing' (Clark, 1968, p.155). Seventeen of the 51 estates were 'idled largely as a result of the political situation created by the Revolution ... for varying periods of time': some remained uncultivated. *Campesinos* had offered a variety of reasons for this loss of productivity e.g. they had obeyed orders from syndicate leaders or 'from La Paz', or they were awaiting a decision on which lands were to be expropriated. On some estates landlords and their former *colonos* had practised a system of share-cropping for several years, until either the *campesinos* had decided to break the agreement or the estates were expropriated. Surprisingly, Clark records that 19 of the estates continuing cultivation at the time of agrarian reform, maintained the flow of farm produce to the city: 'peasants took the landlords' share of production (50 per cent) to the Ministry of Rural Affairs in La Paz for varying periods of time after April 30 1953' i.e. the date on which a supreme decree compelled syndicates to assume responsibility for that year's harvest and for subsequent planting and harvesting of crops.

The process of land redistribution in the study area

As a *comunidad originaria*, which had managed to survive intact, Llamacachi was not directly affected by the process of *dotación*. Unlike small communities, such as Pacharía, which brought successful legal actions against *Hacienda* Chua, Llamacachi was not justified in making any such claim and continued to enjoy free access to grazing lands in Compi. Whereas in 1952 the five *estancias* of Compi formed a joint *campesino* syndicate (which fragmented as a result of in-fighting even before the finalizing of Compi's land adjudication in 1957), there was obviously not the same need for the freeholding to create a peasant union hence, according to residents, no attempts had ever been made to establish such a body. In the mid-1960s eight *campesinos* organized a loose-knit cooperative but this rapidly disbanded once its aims - the purchase of pedigree sheep and improved seeds from the Belén agricultural station - had been achieved.

Although, unlike Chua Visalaya, Llamacachi has not over the years been plagued by countless conflicts arising from the implementation of land reform, redistribution of land in adjacent Compi did indirectly precipitate a bitter feud between Llamacachi's *comunarios* and the *ex-colonos* of Compi. The sequence of events has been recorded in detail by the Buechlers. Whilst in 1957 *Hacienda* Compi's land north of the through-road was categorized and consequently redistributed to the *ex-colonos*, the more productive land between the lake and the road was designated a medium-sized property and thus not subject to expropriation. Because the heirs of the last *patrón* found it virtually impossible to pay the statutory wages to former *colonos*, they opted to form a cooperative with a group of them. Having decided that this arrangement was equally unsatisfactory and failing to persuade *ex-colonos* to purchase the land, the owners eventually sold their fields in 1961 to the eagerly awaiting Llamacacheños, some of them prepared to buy land for as much as 2,000 bolivianos per hectare.

Inter-community hostilities reached a climax in 1965 when *comunarios*, whether or not intentionally, planted crops across a track traditionally used by Compeños, driving cattle and sheep to the lake. So violent were the recriminations that *comunarios* eventually enlisted official aid from Achacachi; on one occasion at least 10 soldiers forcibly detained all Compi's household heads. The question was taken up in La Paz by the CNRA, the prefecture and the Ministry of Peasant Affairs. Finally, finding the cost of legal action increasingly weighty and anxious to restore peace (especially as many *comunarios* were related through marriage to *ex-colonos* in Compi), Llamacacheños agreed resentfully to sell back the newly acquired land.

Other strategies for extending the area of land under cultivation were more successful. *Comunarios* whose spouses had previously been *colonos* living in neighbouring communities benefited from the expropriation and subsequent redistribution of *hacienda* land. In the early 1970s 10 families were known to be renting plots of land on an annual basis from *ex-colonos* in Compi. *Anticrético* (p.116) was more widely practised than in pre-reform times; it became, apart from marriage and inheritance, the most important means of acquiring entitlement to farm plots of land in local communities. Rural-urban migration, which accelerated rapidly in the post-reform period, partly as a result of regular weekend marketing visits to La Paz, contributed in large measure to significant changes in traditional inheritance patterns. By the mid-1960s there was a marked tendency for fathers to transfer property rights to their children prematurely, in an effort to dissuade them from migrating permanently. One *campesino* remarked, 'I did have four hectares but I gave the three boys a hectare each. They're good sons and help me farm my land too'.

Llamacacheños interviewed in 1971 complained that agrarian reform had done nothing to tackle the acute problems of *minifundismo* in freeholdings. Land disputes - especially over damage to property, removal of boundary markers, multiple claims to individual plots and claims to property by illegitimate children -were said to occupy more time at community meetings than any other aspect of community life, with the possible exception of education. A number of households were at the time immersed in legal battles over land; one case had been proceeding for 39 years with no prospect of an acceptable solution in sight.

The particular problems faced by widows and the degree of land fragmentation were reflected in many of the comments volunteered by *campesinos*. 'I have only two small parcels of land - in all less than half a hectare. My husband's land was split amongst the four children ... that left me with just seven furrows'. 'I have three boys and one girl ... we plant the fields together and they all take crops from eight furrows'. 'We only own one plot of land ... my father had very little land and it had to be split amongst five boys and three girls. My uncle allows me to take the produce from a few furrows, if I help him with the work'. 'We have nine parcels of land ... I should say about three hectares - less than my father because

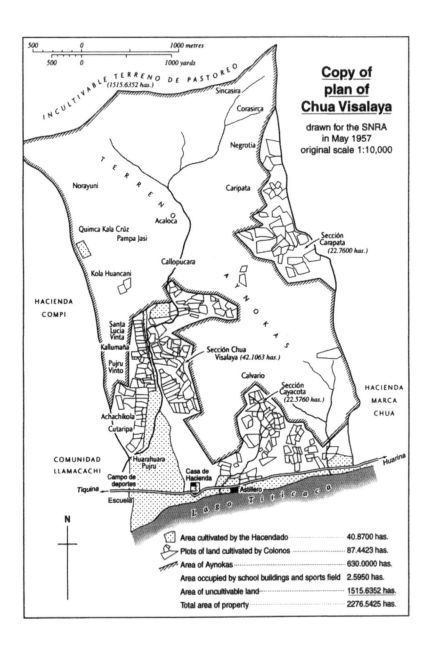

Copy of plan of Chua Visalaya

drawn for the SNRA
in May 1957
original scale 1:10,000

500   0   1000 metres
500   0   1000 yards

INCULTIVABLE TERRENO DE PASTOREO
(1515.6352 has.)

Sincasira

Corasirca

Negrotia

Norayuni

TERRENO

Acaloca

Caripata

Quimca Kala Crúz
Pampa Jasi

Sección
Carapata
(22.7600 has.)

Kola Huancani

Callopucara

A Y N O K A S

HACIENDA
COMPI

Santa
Lucia
Vinta

Kallumaña

Sección Chua
Visalaya (42.1063 has.)

Pujru
Vinto

Calvario

Sección
Cayacota
(22.5760 has.)

HACIENDA
MARCA
CHUA

Achachikola
Cutaripa

COMUNIDAD
LLAMACACHI

Huarahuara
Pujru

Huarina

Campo de
deportes

Casa de
Hacienda

Tiquina

Astillero

Escuela

Lago Titicaca

N

Area cultivated by the Hacendado ················· 40.8700 has.
Plots of land cultivated by Colonos ················· 87.4423 has.
Area of Aynokas ················· 630.0000 has.
Area occupied by school buildings and sports field 2.5950 has.
Area of uncultivable land ················· 1515.6352 has.
Total area of property ················· 2276.5425 has.

136

his land was divided between me and my brother'. A more fortunate farmer divulged that he had been able to buy *hacienda* land: 'I have a half hectare here, which belonged to my mother, and about the same amount of land from my father, a Compeño. I was able to buy about three hectares from *Hacienda* Compi, because I had worked for the *patrón* in the olden days'.

Whilst Agrarian Reform Commissioners had foreseen some of the more obvious situations likely to frustrate and prolong *dotación* procedures, few could have anticipated the variety and complexity of problems presented in due course by a number of communities, almost invariably located in areas of acute population pressure on available land resources. In terms of utter confusion, bitter disputes over plots of land and incidents of violence - culminating during the writer's visit in 1971 in armed combat between rival syndicate and cooperative - Chua could match any community. Few *ex-haciendas* bombarded the CNRA with more petitions and counter accusations; rarely were individual communities the subject of so many official visits and commissions. Chua's drawn-out process of 'land redistribution' clearly illustrates a significant number of shortcomings and persistent problems associated with *dotación*.

Chua Visalaya's peasant syndicate was formed in the early days of the National Revolution, at the instigation of visiting MNR party workers. All Visalayan families remained loyal to the group, which thus provided an unchallenged link between community and government authorities until the creation of the rival cooperative (*Cooperativa Agricola 'Chua' Limitada*) in 1964. Whilst land invasions were rampant in the Lake Titicaca region between April 1952 and August 1953, in Chua, surprisingly, there was no attempt to seize *hacienda* lands: according to the *campesinos*, they had seen no reason to oust the *patrón* since they respected him both as a landlord and good farmer. After the passing of the Agrarian Reform Law, Visalayans assumed the *de facto* ownership of the *sayañas* traditionally farmed by their families (this in fact implied no radical change in land tenure structures since landlords had rarely interfered with inheritance rights to *sayaña* plots). As previously noted, the *patrón* wisely introduced remuneration for estate labour without delay, 'in strict compliance with the law'.

In 1954 the syndicate drafted preliminary petitions for *dotación* with the assistance of the Ministry of Peasant Affairs, and shortly afterwards the property was surveyed by SNRA employees, who reported that 'the greatest obstacle ... arises from disruptions by *campesinos*'. The following year two topographers were commissioned by the SNRA to survey *Hacienda* Chua and produce a plan complete with boundary lines and on which *hacienda* lands, *aynokas* and individual *sayaña* plots were to be clearly delimited. Whilst the topographers established the total area of the Chua Visalaya/Cocani estate and the total number of *campesinos* over 18, they failed to complete the specified plan - mainly because of deliberate *campesino* obstruction, multiple claims to *sayaña* plots and difficulties arising from names (two thirds of Chua Visalaya's claimants had Mamani as their first or second family name).

It was proposed that 2543.7350 hectares should ultimately be made available to the *campesinos* in question; theoretically each *ex-colono* was to be allocated 4 hectares of cultivable land. Ancestral *sayañas* were plotted on the SNRA plan eventually completed in 1957; it was recognized that fragmentation had proceeded to such an extent that no *dotado* (the recipient of land) could be granted legal title to *sayaña* parcels together exceeding 1.5 hectares. In Visalaya a mere 42.1063 hectares were to be parcelled amongst 110 claimants: furthermore, many of the minute plots were classified as *incultivable* e.g. D.M's holding comprised parcels of '0.4000 h., 0.0300 h. (*incultivable*), 0.0460 h. (*incultivable*) 0.0250 h. (*incultivable*), 0.0360 h. and 0.1050 h.'- i.e. a total of 0.6420 hectares in six scattered plots. At the same time 'parts' of *aynoka* territory were to be subject to 'equal division amongst all peasants', regardless of the infertile, rocky nature of the land concerned.

As previously mentioned, the supreme resolution of 1959 designated the property as an agricultural livestock enterprise. According to the Agricultural Bank's later documents, 1949.1500 hectares of the 5599.7650 were allocated to the landlord - a figure at variance with the resolution's extension of 3101.1700 hectares; discrepancies of this order are largely explained by the fact that *dotación* was never satisfactorily concluded within Chua Cocani. The *expediente* contains various letters dated 1967, 1968 and 1970, in which the surveyors previously involved in the complex process of assessment gave numerous reasons for their suspension of work in Cocani. The supreme resolution assigned Chua Visalaya's lands as follows:

Sayañas and *aynokas* belonging to *campesinos*	244.5587 has
Collective *aynoka* lands and pasture lands belonging to *ex-colonos*	722.8836 has
School buildings and sports field	2.5950 has
Hacienda lands in Visalaya	39.3750 has
Rocky area and uncultivable pasture lands	1267.1302 has
Total	2276.5425 has

Thus, as a result of land reform litigation, the *patrón* retained 39 hectares of cultivable land (the community's most productive land) and no *campesino* gained titles to land previously cultivated by the *hacienda*; the resolution merely legalised the *de facto* ownership of fragmented plots of land.

Shortly after the process of *dotación* was completed in 1959, Gordon Barbour, the patrón, drowned in mysterious circumstances: the estate was subsequently mortgaged to the Agricultural Bank, which eventually assumed complete control

138

in 1964. In 1961 the syndicate presented a petition to the CNRA attesting that at least 32 *ex-colonos* had been granted insufficient land i.e. less than 4 hectares, even taking into account uncultivable *aynoka* plots. As late as 1967, six elders of Visalaya (including three syndicate officials) requested a total redistribution of parcels, claiming that some *ex-colonos* had been favoured with cultivable lands whereas others had been granted infertile parcels . When asked to account for such inconsistencies the surveyors concerned replied that they had acted fairly to the best of their ability - they could not be expected to satisfy *ex-colonos'* needs in view of the *superpoblación campesina* (peasant overpopulation): at the same time the *campesinos* received an official reprimand for their refusal to collaborate with SNRA personnel in the 1950s. To the present day, disputes both within and between families over legal ownership of *sayaña* and *aynoka* plots remain unresolved in some cases. Letters from the Federation of Peasant Workers headed 'the land belongs to the man who works it' have never offered consolation to *campesinos* living in a community where for some members there has never been land to work, let alone own.

In theory, this distressing situation should have been alleviated by the transference to the community in 1964 of the 39 hectares of highly-prized lands formerly exploited by the *hacendado*. Had these fields been equally subdivided amongst all *ex-colonos* over the age of 18 (regardless of the problems associated with *minifundismo*) or had they passed collectively into the hands of the entire community, further clashes might have been prevented. Unfortunately, the acquisition of land by 32 *ex-colonos* spelt disaster for the community as a whole: bearing in mind the inter-family friction, mistrust and frequent interruptions to work etc. it is doubtful whether even the privileged cooperative members derived more than marginal benefits. Failing to find a single purchaser, the Agricultural Bank agreed in 1964 to sell the property to the newly formed Chua cooperative (with 32 members in Chua Visalaya and 108 in Chua Cocani) for a sum of 500,000 bolivianos. Although about 50 Visalayans originally expressed an interest in the bank's proposals, the down payment of 300 bolivianos reduced the eventual membership numbers.

The contract stated that until the year 1974 (by which date the debt would be fully paid), bank officials would be entitled to make visits to the property, to offer technical advice and to authorize the sale of crops and livestock. Land, buildings, stock and machinery were legally transferred to the cooperative with the proviso that the fields adjacent to the lake were to remain intact: in fact, in the late 1960s this tract of land between what had been the *hacienda* house and the lake was split into 32 plots, each measuring 12 furrows x 12 metres - on which cooperative families were encouraged to grow their own selection of crops; titles to these plots were confirmed in 1970.

Membership of the cooperative did not result directly in expulsion from the community's syndicate , but in the circumstances it was inevitable that the majority

139

of Visalayans excluded by abject poverty from the cooperative and hence from access to *hacienda* lands to which they considered they had equal claim, should become extremely embittered; open hostility made it expedient for cooperative members to resign *en masse* from the syndicate.

In subsequent years rivalries between the two *campesino* groups were to intensify. By June 1971 litigation was in process on two counts. Syndicate members had brought an action against the cooperative for refusing to allow their families to obtain *totora* from the lakeside; simultaneously they were petitioning the CNRA to prohibit the cooperative from selling eucalyptus trunks to Mina Matilda (a zinc mine) for use as pit props. In the ensuing months frequent accusations were made by members of both factions concerning the stealing of money, the removal of crops from the ground and the poisoning of animals. Early in August fighting broke out after cooperative members resorted to physical force in an effort to prevent syndicate members from walking across cooperative lands to the community's *totora* beds. In mid-August the syndicate's deliberate planting of broad beans on newly ploughed cooperative land precipitated six hours of community warfare (in which sticks, stones, rocks and guns were brought into action), only halted after 15 community members were seriously injured. Although police officials from Achacachi and Huatajata subsequently visited Chua Visalaya, disarmed the *campesinos* and ordered them not to engage in further affrays on pain of a 2,000 bolivianos fine or imprisonment, violence did erupt on a number of occasions in later months.

Whilst considering themselves inextricably entangled in 'the land struggle', *ex-colonos* clearly realized the gravity of the situation and deplored it. The remarks of a syndicate elder recorded at a community meeting in 1971 reflected the sentiments of many *ex-colonos*:

> Members of the cooperative allow me to say that I am your brother because I grew up with you all .... our fathers almost died working for the *hacienda*, receiving the lashes of the *mayordomo's* whip. Why must we fight these days? We belong to one *estancia* and were born on the same land. Fighting could have serious consequences and there is every reason for our fellow-*campesinos* in other communities to criticize us .... I only call for unity.

Other members commented that so-called agrarian reform had done nothing to relieve or resolve old disputes over landownership; on the contrary, it had perpetuated them. The land struggle had become a fight for 'land or starvation - life or death'.

It was virtually impossible to establish the complete landownership pattern in Chua - or in any other lakeside community - at the time. Whilst a few individuals readily divulged the size and location of their *parcelas*, others specified the total number of plots but were reticent about identifying them on the ground: some insisted that the land immediately surrounding their *adobe* dwellings and outhouses represented their entire holdings; many flatly denied ownership of certain plots (known by the field assistant to belong to them) or remarked vaguely that they held 'plots on the hillsides' but said they were frightened to say more in case their neighbours robbed their crops (a common occurrence). A minority admitted that all such land declarations were 'suspect'; *campesinos* were reluctant to discuss ownership because of the ever-present threat of the imposition of a land tax. Three years previously such a tax had been given government approval and only shelved as a result of armed action by lakeside peasant farmers: 'in Belén and Achacachi ... President Barrientos himself had to flee to avoid violence from a peasant assembly which the President had addressed to explain the need for the tax' (Huizer, 1973, p.60).

Landholding patterns were - and still are - complicated by inheritance, marriage and migration. For example, the families of several Visalayan widows whose husbands had been born in Llamacachi retained access to certain plots in the community. Most migrants to La Paz, or its satellite settlement of El Alto, had inherited parental plots (or several furrows in each of a number of fields) and either worked them personally or visited their home community at harvest time to collect the produce due to them: instead of paying family members for field work, they provided free lodging during city visits or bought gifts of sugar, rice etc. Many *ex-colonos* were adamant that their forbears had enjoyed usufruct rights to larger tracts of land than they themselves owned in 1971; land had been further fragmented by inheritance practices. No less significant had been post-reform trends in house building. Whereas in *hacienda* times housing had been confined to rugged, infertile land, during the early 1960s *campesinos* began constructing larger dwellings on valley-side *sayañas*, in some cases drastically reducing the amount of land available for cultivation (in addition to removing topsoil in adjacent plots for making *adobe* bricks).

Some individuals had resorted to extreme measures to acquire land. In Chua Visalaya, a man of 22 confessed to marrying a 40 years-old widow with three children, primarily because she had 'many plots of land', whilst a girl of 18 spoke thus of her 64 years-old husband: 'My mother forced me to marry him because he is an old man with lots of land .... about 4 hectares ... I never even liked him and now I hate him and want to go and live in La Paz'. Whilst few examples of land renting were encountered, land usurpation was relatively common: a number of *campesinos* were convinced that large families provided the sole effective safeguard against the actions of unscrupulous neighbours. One old Visalayan commented: 'My family farmed this land in *hacienda* times ... afterwards a man came from Compi ... he had lived up the hillside with his grandfather. He claimed that the land belonged to him ... I was too old to make much trouble so we split the land. God will see that he is punished in the end'. A desperate shortage of land had provoked seemingly inhuman deeds: a number of old couples and widows alleged that neighbours - in some cases, even their own children - had 'stolen' their land and crops because they were too feeble to prevent them from doing so.

Almost two decades after agrarian reform had held out the prospect of finally resolving 'the land question', inter-community land feuds and landownership disagreements both within and between *campesino* families, continued to dominate and dislocate community life throughout Bolivia's Lake Titicaca region. According to the Agrarian Reform Decree (see p.55), in *ex-haciendas* where land resources were inadequate to create a sufficient number of viable farm units, *campesinos* were entitled to petition for land grants in 'other available areas': as must have been apparent to the Commissioners in 1953, there were no 'other' potentially cultivable 'available areas' within easy access of the lakeside communities.

Colonization

The Agrarian Reform Law had extolled the virtues of opening up the Oriente for commercial farming and designating colonization areas for land-hungry *campesinos*. 'Colonisation of the Amazonian lowland could relieve the population pressure, absorb much of the future population increase, create employment opportunities, reduce food imports and raise exports' (Zoomers, 1997, p.62). Not for the first time in the republic's history was colonization 'regarded as the panacea for the ills of rural Bolivia' (Clark, 1970, p.37). Many politicians and economists. whilst appreciating the urgent need to improve communications between the Yungas and eastern plains, genuinely believed that the mere **existence** of virgin territory awaiting exploitation would be sufficient inducement to mobilize land-needy Altiplano *campesinos*.

According to Article 91 (p.60), all *campesinos* from the Altiplano and Valles zones were legally entitled to 50 hectares of cultivable land in the Oriente region - where 'estate holders tended to welcome *campesino* farming of unclaimed land because they could then tap this labor reserve' and where 'the land reform of the 1950s was largely irrelevant' (Thiesenhusen, 1995, p. 61). Optimistically, the Bolivian government declared its intentions of settling some 100,000 *campesino* families in colonization zones in the Yungas and Oriente during the decade 1962-71. By the end of the period only 30,000 peasant families, most of them 'spontaneous' settlers, had accepted the challenge.

Densely populated Compi, with a well established tradition of migration, was the only community within the study area to have responded enthusiastically to the government's open offer of agricultural land grants and a new life in what was to most lakeside dwellers, an unfamiliar environment. According to reliable sources, 10 Compeño families had taken up land in the Yungas in the late 1960s; during the first year several of the settlers had succumbed to fatal diseases, especially 'the shivering disease' (malaria), and by 1971 others were apparently on the verge of returning to the lakeside or were seeking non-agricultural work in La Paz. A number of Llamacacheños volunteered the information that:'many of the families who went from Achacachi have come back...they couldn't get used to the farming and living there'.

Although one *campesino* from Llamacachi expressed a vague inclination to settle east of La Paz - 'I would like to live near Coroico [the nearest Yungas township] because life is hard here and the work would be easier there' - seemingly not a single *comunario* had taken up a land grant in any of the colonization zones designated in the 1960s. When asked why they had been reluctant to do so, a typical answer was: 'Why should we want to go? The old folk don't want to leave Llamacachi and young people today would rather go and work in La Paz'. In the previous year two Visalayans had migrated to towns in the Yungas, in both cases as a means of securing anonymity. One man had deserted his wife and seven children, 're-married' in the city and was understood to be living with his new wife from La Paz in Chulumani (the largest town in the Yungas), whilst a younger *campesino* was said to have eloped with a Cayacotan girl and intended to remain in the Yungas until his father's wrath abated. Apart from the communities' butchers, buying meat in periodic markets, few Visalayans or Llamacacheños had even visited the Yungas. Several claimed they feared to do so having heard tales of trucks overturning on the hazardous journey from La Paz - still a frequent occurrence on a tortuous, ill-maintained road (referred to as 'the road of death'), which descends precipitously from a 4,725 m pass north-east of the city to about 1,300 m in the environs of Chulumani.

Clearly, the promise alone of readily available, uncontested cultivable land had provided insufficient incentive for land-hungry lakesiders to move to colonization zones. The government's migration and settlement policy had proved incapable of stemming the tide of migration to the *barrios* or *zonas* of La Paz. According to Paz Estenssoro, Chua Visalaya and Llamacachi were 'isolated cases' ... the mass of highland communities had welcomed opportunities for relieving pressure on land and making a new start: no foundation for such a viewpoint was apparent in the lakeside region.

Whilst some politicians and land economists in the 1970s categorically attributed the relative failure of colonization schemes to Andean Indians' inability to acclimatize to tropical lowland environmental conditions, the lakeside *campesinos* interviewed in 1971 dismissed such claims: they pointed out that Aymara communities had flourished in coastal locations in pre-colonial times. They did, however, acknowledge that they were concerned about lowland diseases (especially malaria and Chagas disease, for which there is no known cure) and the lack of a health programme enabling colonists to receive protection where possible from 'new diseases'.

Campesino reasons for not settling in colonization zones were found to be numerous and varied in nature, the overwhelming one being the fact that they had never, as individuals or groups, been adequately consulted by agrarian reform personnel about their feelings and reservations. A general lack of information contributed to their reluctance to migrate eastwards; they had been given no opportunity to make investigative group visits to colonization zones. A deeply engrained attachment to their homeland - the lakeside and the Altiplano generally - was a formidable negative factor: if it had been possible to travel to a family gathering or a lakeside *fiesta* and back to the settlement area within a day, the prospect of migration to the Yungas might not have been so daunting. Some had been concerned that the land they inherited would be 'stolen' in their absence: others said they might have been prepared to re-settle with a group of families known to them, but not as an individual family unit. Having to adjust to cultivating an entirely unfamiliar range of crops (tropical and semi-tropical as distinct from traditional temperate crops) was seen as a major problem. There was a widespread fear of unfamiliar dangerous wild animals - in addition to 'unfriendly spirits', inhabiting the mountains and forests. Not surprisingly, any discussion of colonization focused on the Yungas; *campesinos* within the area regarded the lowland plains of the Oriente as far too remote to have any relevance for them.

Agrarian reform

Whilst textual references in the 1953 Agrarian Reform Decree to the specifically 'agrarian' components of reform were few in number, those objectives that **were** referred to, for example, 'the modernization of agriculture' and 'the permanent transformation of cultivation methods', were radical and extremely ambitious: moreover, they applied equally to *ex-haciendas* and *comunidades originarias*. Were the promises of 'credit facilities and agricultural extension', 'advice and technical assistance', 'agricultural mechanization' and 'rural farm shops ... selling seeds, fertilizers, chemical substances, machinery and implements' etc. fulfilled? If so, did *ex-haciendas* and freeholdings take advantage to the same extent of what was on offer from the Agricultural Bank, research stations, extension agents and new marketing facilities? Did the changes introduced in the 1950s and 1960s 'stimulate greater productivity' (the fourth 'fundamental objective'), thereby satisfying one of the MNR's principal aims of improving efficiency?

Inevitably, Altiplano communities lying off the beaten track, far distant from urban marketing centres and exposed to the harshness of the climate were severely handicapped in their efforts, however well-intentioned and praiseworthy, to adjust to the post-*hacienda* situation: government representatives were reluctant to visit them and consequently they were largely denied access to the benefits of agrarian reform. For many *campesinos,* migration, rather than agricultural development and increased participation in marketing activities, offered the only feasible means of advancement.

According to Clark (1970), 'the area in which most significant changes were produced lies within a radius of 4 to 6 travel hours from La Paz'. Lakeside communities, such as Llamacachi, Chua, Compi and Huatajata, all situated within 120 km of La Paz, the country's largest marketing centre, were ideally positioned to exploit their proximity to the city and favourable climate. As previously mentioned, the Copacabana-Huarina through-road (a route originally avoided by merchants and travellers) had been improved in the 1940s: in 1963, in preparation for a national car rally, the road was further widened and formidable potholes filled. Thus by the mid-1960s it was possible for lakeside *campesinos* to deliver market produce to La Paz and complete the return journey by nightfall; likewise CNRA and Agricultural Bank personnel could visit specific communities within the course of a day. Copacabana's growing popularity, both as a lakeside resort and a pilgrimage centre, Compi's annual folklore festival (first held in 1965 and including reed boat races and the crowning of the Potato Queen) and Huatajata's thriving yacht club all helped to put the lakeside communities on the map and guaranteed the maintenance of the lake road, vital to successful marketing.

It was abundantly clear in 1971 that the study area had been, and still was, well served by its regional agricultural research centre and extension agents (TDCs) or village level workers in Peace Corps parlance. The Belén experimental station, two miles north-west of Achacachi, together with three other research centres, had been established as a result of the formation in 1948 of the Inter-American Agricultural Service, a joint undertaking by the Bolivian and US governments to provide assistance in agricultural research, extension, mechanization and credit. Belén had originally been intended to concentrate experimentation on sheep and potatoes. Since its foundation, an international team of agricultural experts had carried out tests on different breeds of animals; by the late 1960s they had achieved creditable results with Holstein and Brown Swiss cattle, with Romney Marsh and Corriedale sheep and New Hampshire hens. Chua's last *patrón* had from 1948 to 1959 purchased livestock from Belén and participated in crop and animal husbandry experiments supervised by the station's staff. From 1968 Belén also functioned as one of six farm institutes, offering four year practical courses to *campesinos* in La Paz department.

From 1964, the year after the research station was extended, Belén also incorporated the National Community Development Programme's central office for the four provinces of Omasuyos, Los Andes, Manco Kapac and Larecaja. According to Baumann (1970), the role of the NCDP, devised by Bolivians and Peace Corps staff, was to cater for 'the felt needs' of *campesinos* -'to develop strong, self-reliant rural communities capable of identifying and solving their problems through the promotion of local community organizations'. Each TDC was expected to complete the following tasks:

> to stimulate the villagers to become aware of their problems, express the needs they feel, teach them the practical material that he has been taught, and help them to make effective use of the existing technical services at their disposal through specialists who are assigned to work at the provincial level in the area operational offices.

By the early 1970s, 22 NCDP members of staff (18 TDCs and 4 supervisors) were operating in the four provinces. All of them had received training at Belén and remained in close contact with the NCDP regional office. The TDC working in the Chua/Llamacachi area, a returned Llamacacheño migrant, had been responsible for a number of important innovations in the communities viz. the vaccination of creole cattle, the purchase by several Llamacacheños of 30 pedigree sheep and sacks of seed potatoes from Belén, the adoption of certain inorganic fertilizers and the fumigation of stagnant water.

Whilst lakeside communities were being visited on a regular basis by TDCs, individual *campesinos* were also able to combine weekly marketing transactions in Huatajata and Jank'o Amaya with visits to NCDP-managed stores, where they could seek advice on farming problems and purchase a wide range of improved seeds, fertilizers and insecticides. Illustrated pamphlets and booklets on a variety of farming topics, such as how to apply fertilizers and prevent crop diseases, were also readily available. It was normal practice for Belén and the NCDP to mount a joint exhibition on 'new' crops and farming techniques at Compi's folklore festival.

The fourth objective of the Agrarian Reform Decree had emphasized the need for 'encouraging the investment of new capital' and 'opening possibilities for credit'. To what extent did individual *campesinos*, syndicates and cooperatives have access to credit facilities in the post-reform period? Burke's late 1960s investigations led him to conclude that there had been 'no inflow of agricultural equipment in the area [the lakeside region] either for replacement or for addition to stock since 1953, because the Bolivian *campesinos* have neither the funds nor the inclination to purchase this type of capital'. It could be argued that the older generation of *campesinos* preferred to retain small amounts of surplus cash in 'ceremonial funds' and invest such scarce financial resources in *fiestas* rather than farming and that the concept of borrowing large sums of money from an impersonal institution or pledging land and livestock as a surety was repugnant - but, unfortunately, this hypothesis was rarely put to the test in the lakeside area. The credit facilities alluded to in the agrarian reform decree were not forthcoming, at least not where the *campesino* was concerned. García (1970, p. 334) observed that: 'Both the Central Bank and the Bolivian Agricultural Bank have been averse to the establishment of any kind of credit system for the rural cooperatives, or for peasants who have received land through agrarian reform'. Whilst the Agricultural Bank was prepared to support the large agricultural enterprises (including the sugar mills) in the Oriente, it was generally reluctant 'to accept as collateral the titles issued by the Agrarian Reform Service' (García), arguing that the difficulties and costs of collecting small loans would exceed the chargeable interest of 12 per cent.

This being so, *Cooperativa Chua Limitada's* relationship with the Agricultural Bank was anomalous; the bank only resorted to selling the *empresa agricola* to the cooperative after failing to interest a 'suitable' landowning client. Whilst the cooperative was able to purchase improved potatoes and livestock medicines from the bank at reduced prices - reasoning that this would help members to fulfil their financial obligations - the bank's mortgage arrangements placed a considerable financial strain on the community.

García affirmed that, 'institutional agricultural credit has not had any part in the development of the rural communities which benefited from agrarian reform'. Not only did the majority of *campesinos* lack the necessary cash resources for purchasing seeds, animals, fertilizers, insecticides, tools etc., they were also denied access to any crop insurance policy ensuring future credit provision in the event of crop failure from plant disease or climatic 'plagues' viz. flooding, freak gales or hailstorms. They were obliged to rely exclusively on small loans from more fortunate family members, remittances from migrant children or forward sales e.g. by 1971 it was commonplace for onion vendors from the lakeside to buy crops per irrigation unit and pay the producer on return from the marketing transactions.

Clark (1970, p. 61) has attributed 'the "apparent" decline in agricultural production after 1952' largely to 'marketing adjustments and transportation scarcity'. According to him, 'the peasants, their unions and leaders, as well as local officials and middlemen, responded to the bottleneck in marketing and transportation which had been created by the Land Reform by creating a large number of new fairs and by rapidly increasing the number of trucks visiting these areas'. Certainly, lakeside-dwelling *campesinos* interviewed in 1971 reported that the volume of traffic had increased at an unprecedented rate in response to ever-expanding marketing activities, especially after the widening of the road in 1963. By 1971 as many as 50 trucks, transporting traders and produce, passed through the area on local market days. (Bolivian and Peruvian bus companies at that time normally refused to carry heavily-laden *campesinos* for short distances).

On the other hand, the main response to Clark's 'bottleneck in marketing' had not been a rapid growth in the actual number of rural periodic markets but rather a spontaneous expansion in the size and scope of pre-existing ones. Buechler's analysis, *Peasant Marketing and Social Revolution in the State of La Paz, Bolivia* (1972), recorded that in the late 1960s some 65 weekly, 23 annual and 'several hundred festive fairs' were being held in the department. None of the annual events and only 11 of the 65 weekly fairs had been inaugurated since 1953. In Omasuyos, the markets of Jank'o Amaya (Thursday), Huarina (Sunday) and Achacachi (Sunday) had all originated in the pre-reform era; Huatajata's thriving Wednesday fair was the sole surviving post-reform addition. In several communities well-intentioned syndicates or groups of *campesinos* had been forced to abandon attempts to establish new markets because of their failure to attract sufficient numbers of traders. Soncachi provided the perfect example of an incipient market doomed to fail; it was less than 5 km from Huatajata and, in any case, lacked a central plaza adjacent to the main road. Buechler visiting the markets of Huatajata and Jank'o Amaya in 1969 was impressed by the 'unusually high percentage of rurally-based traders' frequenting them - 76 per cent of 339 vendors at Jank'o Amaya and 89 per cent of 227 at Huatajata. This she attributed partly to 'the general population density and long history of trading' in the lakeside region but equally to the ready availability of Altiplano staples (potatoes, *chuño, oca, quinua, cañahua* and dried beans) and contraband goods from Peru.

Throughout the 1950s and 1960s city markets had expanded significantly in terms of both size and facilities: new markets had developed on sites adjacent to the pre-reform foundations 'in primary, older commercial and residential zones and in older well-established suburbs' (Buechler, 1972), whilst more than a dozen had taken root in migrant *zonas* in La Paz and El Alto. Additionally, the central government had authorized the setting up of some 20 *tambos* (open markets serving as distribution centres for city markets) enabling truckers from rural areas to deliver their loads without interference from tax collectors or market police. Moreover, whereas in *hacienda* times *hacendados* and *mestizo* middlemen had virtually monopolized city marketing transactions, not infrequently unscrupulously exploiting rural communities, by the mid-1960s *campesinas* were demonstrating their skills as entrepreneurs intent on maximizing their families' incomes by assuming total responsibility for the direct sale of farm produce to customers in La Paz, thereby dispensing with the costly services of city-born middlemen.

9 *Campesino* opinions on agrarian reform

By 1971, less than 20 years after the introduction of agrarian reform, very little tangible evidence of the *hacienda* system remained in the study region. *Hacienda*-built stone walls and a few scattered eucalyptus plantations were totally eclipsed by the amazing concentrations of pre-Columbian terracing around the shores of the lake. Several *hacienda* houses had survived and been up-graded. The Huatajatan Baptist mission was still working part of the original estate but under constant pressure from community elders to relinquish control of land in favour of resident *campesinos*. Elsewhere *campesino* farming dominated the landscape.

The comments on agrarian reform and rural change recorded in this chapter are drawn largely from the contiguous lakeside communities of Llamacachi and Chua Visalaya. It could, of course, be argued that the eventful histories of these two communities make them both 'unique' - yet every farming community is unique in some respect. The opinions expressed by Chua Visalayans and Llamacacheños in 1971 and 1981 reflected a wide range of age, outlook, income and work experience.

By the early 1970s the contrasts between the two communities, in terms of agricultural productivity and marketing activities, were striking. Whilst the Visalayans' obligations to the *hacienda* had ceased in 1953, the Agricultural Bank's land settlement with the community's better endowed *campesino* families had aggravated already deep seated conflicts over landownership. Fragmentation and landlessness posed ever-increasing problems for the majority of farmers who continued to live at bare subsistence level, farming on a precarious basis. By contrast, neighbouring Llamacacheños, capitalizing on their traditional skills and qualities of initiative, determination and resilience, had successfully managed to exploit the new marketing opportunities made available by the implementation of agrarian reform. By 1971 a lucrative cash crop economy had evolved causing a

149

number of nearby communities to envy the new-found wealth of the traditional freeholding.

Chua Visalaya's experience of agrarian reform

(i) *Cooperative farming*

Formidable problems had reduced the effectiveness of *Cooperativa Chua Limitada*. In its formative years a number of *ex-colonos* had withdrawn their support and membership, being loath to pledge land, livestock and buildings as surety. Some members complained that cooperative tasks were ordered in such a way that they were not free to market their own produce at will, nor to undertake work on a temporary basis in La Paz or elsewhere. Numerous *campesinos* resented the frequent disruptions to field work by the ever-increasing number of meetings: a few even asserted that they were more tied to the community than in the days of the *hacienda*. It was impossible to ascertain with any measure of accuracy the amount of money obtained from crop and livestock sales; some members genuinely appeared to have no idea of the sums involved, whilst others distrusted the cooperative's leaders and were convinced they retained more than their fair share of profits. Although there was no proof of this being the case, certainly the cooperative's potential earnings were being considerably reduced by the fact that produce was being marketed by intermediaries i.e. by entrepreneurs from La Paz.

Whilst cooperative members refused to divulge details about harvests, it was evident that yields from *ex-hacienda* fields in 1971 were significantly lower than those obtained in the last days of the estate - a point frequently made by *campesinos* of neighbouring communities. What was abundantly clear was the fact that Chua's traditional dual economy had disintegrated: all hopes of developing an enterprising cash crop economy had been dashed by the cooperative's debt burden imposed by the bank.

Whereas in the early days of the cooperative sacks of improved potatoes (mainly *Sani Imilla*) had been supplied by the bank or bought from local NCDP farm stores, no new seeds or tubers were being planted in cooperative fields in 1971; as elsewhere in the community, old seeds were sown, regardless of whether or not they were diseased. Chemical fertilizers were only being applied to land assigned for the production of potatoes. All crop experiments appeared to have ceased; visits by TDCs and government representatives were concerned mainly with attempting to resolve conflicts over land etc. rather than normal extension services. Although the Agricultural Bank's agreement had formally transferred all estate machinery to the group, no evidence was found of *hacienda* equipment (except for saws) being used in farming activities; much valuable machinery had been allowed to rot and various components had been removed for sale in La Paz. Cooperative members had reverted to traditional ploughing methods: on one occasion 15 teams of oxen were observed ploughing a patch of land not exceeding half a hectare in extent. Whilst this was understandable in view of fuel and maintenance costs and the abundance of readily-available labour, such animals imposed a burden on the community, consuming valuable fodder, frequently escaping, in so doing causing crop destruction and, not infrequently, bodily injury. Work within the cooperative normally proceeded on a collective basis, all members participating simultaneously. In the example quoted, those unable to supply oxen for ploughing were obliged to work - as were their wives - with digging sticks or clod breakers. Seeds were sown broadcast and all harvesting done by hand.

Numerous syndicate members maintained that, on concluding their negotiations with the bank, cooperative members and their families had squandered money (obtained from the sale of crops and machinery) on lavish *fiestas* and prolonged drinking bouts; they had also slaughtered many of the estate's pedigree animals for their own consumption. Whether or not such allegations were justified, there had been a marked deterioration in both the quality and quantity of stock; by 1971 the cooperative owned only two pedigree cows and a flock of 120 inferior quality sheep, entirely dependent for their sustenance on local grasses.

(ii)   *The persistence of subsistence agriculture*

For the majority of *campesino* families the death of the *patrón* in 1959 had brought few, if any, benefits. True, they acquired access to unlimited time for their own farming tasks and leisure activities - though a number of women alleged that this had been counter-productive, resulting in heavier drinking! On the other hand, the payment they had received regularly since 1953 for estate work ceased forthwith and they gained nothing in terms of land grants from the Agricultural Bank's settlement with cooperative members. For the first time in their lives *campesinos* were forced to make their own decisions about what crops to grow, fallow periods, livestock, marketing etc. but, in view of the total size of their holdings and fragmentation problems, they were compelled to grow what they regarded as vital to family needs.

The potato continued to occupy pride of place throughout the community. Native varieties, such as *sakapamya* and *waca laryara* (low yielding, small and susceptible to several types of blight) were being harvested after Easter and then converted into *chuño* and *tunta* during June (usually the coldest month), before being stored in clamps at the centre of family compounds. Whilst traditional crops, such as *oca, papalisa* and *quinua*, were still being raised, there had been a slight shift in crop emphasis. Broad beans, planted in September and harvested in June, were being cultivated more widely and since the late 1960s a few families had

151

begun to produce onions (albeit on a small scale) in imitation of their neighbours in Llamacachi. Crop diseases were rife because of the planting of infected seeds and tubers or the sowing of unhealthy seeds in contaminated soils. When asked why they had not accepted the advice of the local TDC, *campesinos* replied that the seeds were 'too expensive' or, with a traditional fatalism, that 'you cannot alter the course of nature'.

As throughout the lakeside region, one of the main adjustments in Chua Visalaya's post-reform farming had been an intensification in land usage, reflected in the gradual disappearance of the traditional period *en descanso* (resting period/fallow): this had been accelerated by an unprecedented phase of landholding subdivision. A veritable boom in house building had aggravated the situation by reducing the overall area of land for farming. By 1971 cultivation was continuous on many *sayaña* plots; any land left fallow for several months was being unrestrainedly grazed by livestock. Not surprisingly a number of *campesinos* bemoaned the fact that some of their fields had become virtually sterile and yields were declining year by year. Very few *ex-colonos* outside the cooperative were applying any form of artificial fertilizer; unlike the cooperative members and the majority of Llamacacheños, syndicate members were even denied access to *totora* from the shore.

Whilst most of the *ex-colonos* alleged that, except for llamas, they owned more animals in the early 1970s than they had done in *hacienda* times, livestock averages per family were very low: cattle, 1.5; sheep, 14.9; pigs, 1.8; llamas 1.1; donkeys, 0.2 and poultry, 0.4. Lack of fencing to check the spread of contagious diseases, infected soils, underfeeding (confirmed by the visit of UN personnel) and poor husbandry (e.g. the tying of sheep's hind legs to curb straying, eventually laming many animals) all contributed to the depressing state of livestock farming. Although anti-swine fever doses and pig cholera vaccines were readily available from the TDC, very few *campesinos* were taking any preventative or curative measures, since the majority expected animal losses to occur and accepted them as inevitable.

From the statistics it is clear that the main emphasis in livestock farming was on sheep breeding. Sheep were highly valued; they were presented as dowries, sometimes sold by *campesino* families to finance household or community fiestas, but kept primarily for their wool which, because of its inferior quality, was used for weaving crude sacking and bed covers. It was clear that unrestricted grazing of land by sheep had promoted land degradation by loosening soil particles and exposing topsoils to erosion by strong winds and occasional damaging storms. Llamas, usually bought from Chileans or Peruvians at Rio Seco (the custom post near La Paz) were kept on a temporary basis and freely grazed on the *pampas* when not required for transporting produce; after being employed as beasts of burden during the harvest season, they would be sold to a local butcher, who dried and salted the meat for sale (as *chalona*) at one of the weekly markets in the

Yungas. Hens had been introduced in the 1960s as part of the Peace Corps' regional programme but, according to lakesiders, their numbers had declined disastrously 'because of climatic problems'. There was no evidence of any animals being reared on concentrated foodstuffs.

Such was the general depressing state of agriculture in Chua Visalaya. Several of the better educated young *campesinos*, whilst deprecating the general situation, felt they lacked the financial means and expertise to improve standards. One was genuinely eager to embark on the Belén course in practical farming, but lacked the funds. One progressive *ex-colono* had planted about an acre of land with *alfalfa*, having purchased the seeds from La Paz. Not surprisingly, some of the more go-ahead young *campesinos* were frustrated by the slow rate of progress: the comment, '*no deseo ser hombre del campo*' (I don't want to be a farm worker) was heard on numerous occasions.

The emergence of cash crop farming in Llamacachi

Whilst farming practices in the *ex-hacienda* continued along traditional lines, agriculture had been virtually transformed in Llamacachi during the post-reform period. This is not to imply that vestiges of the past were difficult to find: indigenous crops such as *isañu* and *quinua* were still being grown on *sayaña* plots; farming tasks were still carried out on the specific dates prescribed; some Llamacacheños continued to approach animal losses with a traditional fatalism - 'the sheep always die in February' - and great faith was still placed in the *brujo's* (witchdoctor's) rain-creating and hail-preventing abilities. But for a growing number of *comunarios* agriculture no longer represented a tedious but necessary way of life; it had taken on the aspects of a business enterprise.

The most obvious modification was the replacement of potatoes by onions as the principal crop. As one *comunario* aptly observed, 'before, all was potatoes, *oca* and barley ... now it's all onions, onions and onions'. Time-honoured rotation systems had been dislocated; whilst no Llamacacheño could remember lakeside fields being left fallow for an entire year, unquestionably they were being used more intensively than in pre-reform times. 'Since onions grow well through the winter months, there's no reason why we shouldn't grow them then and catch the market'. Although some farmers insisted that they never produced more than two crops of onions from any plot of land within a six years period, others admitted to finding it convenient and profitable to raise two crops on the same plot during the course of a year; seeds could be sown in nursery beds in January/February, transplanted in March/April and harvested in June, whilst a second crop could be lifted in November/December. With the application of selected fertilizers, careful nurture at all stages of growth, regular irrigation (by means of cans and buckets of water from irrigation ditches or the main stream), high yields could be obtained from lakeside fields in Llamacachi, as was also the case in the communities of Compi and Huatajata.

Despite the emphasis placed on onions, the potato still occupied a significant position in Llamacachi's economy. According to *comunarios*, total yields far exceeded those of pre-reform days: this they attributed to two innovations viz. the application of chemical fertilizers, pesticides and insecticides and the near-total reliance on improved varieties. Various progressive *campesinos* taking advantage of the advice offered by the local TDC or Belén, had experimented with crops not normally included in rotation systems e.g. several had grown *alfalfa* for fodder. One family had cultivated carrots and radishes. The roots had been too small to compete favourably in Paceño markets with those from the Cochabamba area, but the fact that they were actually produced in the first place was indicative of progressive attitudes to farming. A number of *campesinos* were making full use of the advisory materials available in local NCDP stores; some had bought illustrated booklets on how to obtain high yields from potatoes and on disease prevention in sheep and poultry.

Whilst animal losses, especially sheep (72 in the previous year), were still a cause for deep concern in Llamacachi, there had been appreciable advances in the livestock sector. The general consensus of opinion was that stock numbers had increased considerably in the post-reform period and that the quality (even of creole stock), had greatly improved. A few families had bought Corriedale sheep from Belén; in 1971 the community owned 31 pure-bred sheep. Since the mid-1950s farmers had also been buying pedigree pigs from Huatajatan dealers, most of the stock originating from the Cochabamba area. As in Visalaya, importance was attached to purchasing mules or donkeys, highly esteemed as beasts of burden. Some farmers had tried keeping hens but with little success because of the very low hatching percentages and respiratory problems; by 1971 most families had given up the attempt and no household possessed more than eight birds. Unlike Visalayans, an increasing number of Llamacacheños were feeding their animals on supplementary rations; some *campesinos* were buying sacks of wheat flour for this purpose. Most of the farmers were taking advantage of the availability of cheap vaccines, supplied by the TDC.

Although by the late 1960s the bulk of Llamacachi's onions and some home-produced foodstuffs were being sold directly in Paceño *ferias francas* (open markets), most Llamacacheños frequented at least one local weekly fair. There, in company with *ex-colono* neighbours, they would engage in small-scale negotiations, usually involving cash transfers but sometimes barter. However, unlike Visalayan women, Llamacacheñas were operating as skilled entrepreneurs, acting as 'middlemen' in the sale of lakeside produce. For example, two shrewd young women were running a lucrative business in dairy produce alongside their normal onion selling activities, buying eggs and home-made cheeses (against strong competition from other lakeside *campesinas*) for resale in La Paz. The 300 or 400 eggs, bought from a large number of families, were conveyed to La Paz in baskets packed with straw and were sold the following day to regular customers, including a restaurant owner.

All market transactions were dwarfed by the flourishing trade in onions for which Llamacachi and the *estancias* of adjacent Compi had in the previous decade earned renown throughout the lakeside region and the northern Altiplano. The visitor to Llamacachi could not fail to be impressed by the magnitude of post-reform agricultural change, by the initiative, risk-taking and expertise which had made the transition possible and by the sophistication of marketing in the early 1970s. For a community to base a cash crop economy almost entirely on a vegetable hardly grown in the region before the present century might appear hazardous in the extreme; in the event the onion proved an ideal choice. Apart from the favourable climatic conditions, onion production was labour intensive, providing year-round employment for Llamacachi's families and local impecunious *ex-colonos* if the need arose. Proximity to La Paz, Bolivia's largest consumer market, placed lakeside vendors at a clear advantage over Cochabambina rivals, obliged to truck their bulky commodity some 370 km over mountain roads. Moreover, onion marketing exercised the latent ingenuity of *comunarias*: most young women had readily accepted the role of onion vendor convinced that a 'dual' existence - four days in the community followed by three (Friday, Saturday and Sunday) in the city - offered the best of both worlds: at least 13 of Llamacachi's migrants had met their husbands whilst in La Paz primarily for the purpose of selling their families' onions! Whilst marketing itself remained a predominantly female occupation its success obviously depended to a considerable extent on the availability of cheap and regular transport i.e. on the cooperation of the community's truckers.

By 1971 some 40 Llamacacheñas (excluding migrant vendors in La Paz) were making weekly or twice-weekly journeys to sell onions; a further 16 made between five and 10 trips annually. In addition to marketing their own produce, some were purchasing (or obtaining on credit) bundles of onions from neighbours or *campesinos* in nearby lakeside communities. Others had become expert at transacting business in city *tambos:* there they could procure onions for resale from as far afield as southern Peru. By 1971 vendors were operating according to a set routine, transporting onions to the city in one of Llamacachi's two lorries (at reduced rates in return for regular custom) on Friday morning, spending the afternoon visiting relatives or purchasing onions in *tambos,* selling their onions in the Avenida Montes *feria franca* on Saturday and Sunday, eventually returning to the lakeside on Monday to resume household and field tasks. Some of the more competent saleswomen had managed to curtail their marketing hours by building up a regular clientele, partly through the recommendations of city-based relatives but also by retaining their highest quality produce for certain customers, by giving extras or adding a few sweets to each bundle or sack of onions sold. The Avenida

Montes market syndicate established in 1966 had by 1971 become one of the strongest market unions, claiming a membership of 500 female vendors. It boasted its own bilingual radio announcer, broadcasting news of syndicate meetings, *fiestas* and demonstrations, as well as publicizing requests for help on behalf of its members. As a result of the intervention of several forthright Llamacacheñas occupying key posts in the syndicate hierarchy (the author's field assistant had on various occasions acted as chief spokeswoman), for a daily rent of 20 centavos the individual vendor was entitled to display her wares in a prominent section of the market strictly reserved for usage by *campesinas* from Llamacachi and Compi.

La vida linda

'I want only to die ... my husband died last month with a bad cough. I have the same illness and my chest hurts when I cough. There's nothing to eat in the house but I don't want to eat. It's very cold at night but I haven't got the strength to light a fire. My sister's family looks after the donkey but the neighbours have taken all our land because we were too weak to prevent it'. To this elderly Visalayan widow, a number of aged couples, and several desperately poor and infirm *campesinos* interviewed in 1971, life was exceedingly burdensome and the concept of agricultural change largely irrelevant; in fact they were of the opinion that in *hacienda* times people in similar circumstances had received kinder treatment from both younger members of their families and fellow-*colonos* or *comunarios*.

With these notable exceptions, middle-aged and elderly *comunarios* and *ex-colonos* alike i.e. those *campesinos* who had experienced the rigour of life in pre-reform Bolivia and had witnessed the entire process of change in the lakeside region, were unanimous in their praise of 'Taita Paz': 'we will always be indebted to Dr. Paz who gave us our freedom' (the comment of a Visalayan *ex-colono*). Despite the seemingly insoluble land disputes, the majority of *campesinos* insisted that life in 1971 was much pleasanter than it had been 20 years previously: there was frequent reference to *la vida linda* (the fine/nice life). As one 70 years-old *ex-colono* reasoned. 'We're much happier now .. life is not so hard and my grandson helps me in the fields. We suffered a great deal in the past and had to get up at 4 o'clock to go to work for the *patrón* ... Really, the young people don't know what it was like. The children [seven] didn't go to school....not like they all do today. There was never enough food to eat at home and there was no time to make proper clothes for the children'.

Understandably, for many *campesinos* the destruction of the *hacienda* and all its connotations represented the greatest single achievement of the National Revolution. As Erasmus (1967) discovered in south-east Bolivia: 'most peasants preferred to emphasize the elimination of labor obligations to their *patrón*' rather

156

than any other benefits. What remained uppermost in the minds of more than a few forthright *campesinos* was what Heath (1970, p.8) referred to as 'the enormous psychological impact of the revolution' - the acquisition of the intangible personal dignity. Nearly all *campesinos*, irrespective of whether they had been directly subject to the estate system, were eager to point out that, whereas formerly they had been treated as 'donkeys' or 'children' and from lack of education had been reduced to behaving 'like sheep', 'now we're all on the same level ... nobody can tell us what to do any longer'.

As anticipated, informants in both lakeside communities stressed the importance of several or all of the following aspects of change: improved living standards (in terms of food consumption, housing, clothes and material possessions), freedom of movement, the possibilities for greater participation in marketing activities, close contact with La Paz and, not least, education for their children and grandchildren. Whilst agrarian reform had signified no modification in Llamacachi's ancient pattern of land tenure, *comunarios* were in agreement about other benefits. A number had valued the opportunity to gain access, through renting, to additional plots of cultivable land in Compi: the age-old dread of land expropriation by unscrupulous landlords had faded away; instead, some of the more prosperous families were able to employ *ex-colonos* on a casual basis at peak agricultural periods. Farmers had welcomed the advice of TDCs and the extension services offered by the Belén agricultural research station. The expansion of marketing facilities both in the countryside and La Paz had been the greatest boon. The rapid development and prosperity of the onion trade was specified by a number of Llamacacheños as the prime indicator of community progress.

Although feelings about modifications in traditional life styles were generally favourable, rarely did one encounter an elderly *campesino* who believed that all change had been for the better: on the contrary, certain aspects of contemporary life were sorely deprecated. Understandably Chua Visalayans, whilst fervently expressing their gratitude to the MNR for liberating them from the shackles of the *colonato* system, bemoaned the fact that land redistribution had in no sense resolved the problems of landlessness and fragmentation; instead it had provoked unparalleled bitterness and mistrust between *ex-colono* families, culminating in violent clashes and serious bodily injury. Frequently *campesinas* commented that their husbands seemed to have 'lost the will to work'; they had grown lazy and were spending far too much time on meetings and brawling with members of the rival faction.

Whilst the majority of Llamacachi's farmers emphasized the obvious advantages of being able to procure better seeds, chemical fertilizers, pesticides etc., as a means of raising output, a number reluctantly acknowledged that the ever-accelerating pace of rural-urban migration was presenting them with considerable labour difficulties. They were adamant that land hunger had been the root cause of migration 'in the early days' (pre-Revolution times). Eight of the householders interviewed insisted that in the late 1960s members of their families (in two cases as many as four children were involved) had been compelled to leave the community 'from economic necessity' - because of *'falta de terreno'* and *'falta de plata'* (lack of land and money). The following are typical of the reasons given: 'We're farmers at heart but there's very little land in our family and not enough work to occupy us all ... so they had to find work away from Llamacachi'; 'The two boys went to La Paz because there's nothing for them here ... we only own about half a hectare' and 'My four sons went to the city because I only had a few small *parcelas*...three of the boys married girls in La Paz and the youngest is still there at school'. (Significantly, 24 - 20 boys and four girls - of the total 92 migrant Llamacacheños living in La Paz in 1971 were school children, unlikely to return, except for short visits, to the community of their birth).

Migration from the lakeside had in pre-reform days been regarded as a convenient means of alleviating population pressure on limited land resources: by 1971 it was being recognized as an ever-growing problem by lakeside farming communities. A 64 years-old Llamacacheño, one of the largest landowners in the community, explained the difficulties: 'My big problem is not that I'm short of land but that there's nobody left to help me in the fields, especially when my daughter's selling onions in La Paz. The three boys have all left the community and don't intend to return ... I have to employ men and women from Chua when they're available'.

Young lakeside dwellers' attitudes to farming and rural life differed but, taken together, indicated a very bleak future for farming within the lakeside region. 'I'd quite like to work in La Paz but I'm happy enough here' (a young Llamacacheño at school in Compi). 'After I've finished school in Compi I'm going to have three months rest. I will go and look for a job in La Paz ... not that I really want to leave Llamacachi but there are five of us in the family, we have hardly any land and my father has no money to hand over'. 'There's nothing to do here .... I don't want to be a farm labourer' (a boy from Llamacachi on holiday from school in La Paz). 'I don't want to carry on working on the land ... I'm at school in Compi and when I finish the exams I want to go and live with my brother in the city. Why should I have to struggle to survive here?' (a Visalayan boy). 'I want to go and study to become a lawyer. We're not really happy here in Visalaya because we've got no land worth speaking of. I have to go and work in La Paz in the holidays to support the family. I would like my children to be able to speak Spanish properly and have the chance to get on in life'. Whilst a few young men were hoping to remain in the lakeside communities, the vast majority either felt obliged to migrate because of the lack of land and work opportunities or were eager to move to La Paz as soon as possible. Young Llamacacheños with land expectations but whose brothers had 'made good' in the city were unsettled and particularly vulnerable to the persuasive arguments of city-dwelling kinsfolk, especially at *fiesta* time. In a number of families the sole remaining son was keen to join his brothers in La Paz but was prevented from doing so by domestic circumstances i.e. the obligation to assist aged parents with field tasks, thereby ensuring a continuous supply of onions for marketing. Since all 10 Visalayan female migrants between 13 and 24 were working as almost invariably exploited *domésticas* (domestic servants), to supplement meagre family incomes, such comments as the following were hardly surprising. 'They should build a small textile factory by the lake and give the girls something to do ... the men have their weaving and fishing but what is there for us to do apart from working in the fields and selling onions?'

What did *campesinos* themselves identify as their communities' greatest problems in relationship to farming? With few exceptions, household heads in both communities maintained that 'the land problem' throughout the region and a general lack of community cohesion constituted the overwhelming causes for concern. It was widely acknowledged that communities torn apart by struggles over land were powerless to respond positively to well-intentioned offers of assistance from outside agencies. In Chua Visalaya the confrontations between rival syndicate and cooperative had created total discord and confusion; elders could no longer meet together as an integrated group to discuss the community's needs and problems. In such circumstances, any community development projects, even if initiated, were doomed to failure.

Throughout the pre-reform period, despite long-term inter-family disagreements arising from rival land claims, a general spirit of harmony and cooperation had prevailed in Llamacachi; *comunarios* had been united by their ever-present fear of land seizure by expansionist *hacendados*. Since the early 1950s religion had virtually replaced land as the main cause of feuds and misunderstandings: Baptist elders (accounting for half the community's total) were being accused of undermining Aymara traditions and beliefs and of hoarding money. Baptists, for their part, were accusing fellow-*campesinos* of wasting time and money on *fiestas*: over-drinking had 'prevented a number of *comunarios* from becoming successful farmers and entrepreneurs'.

Undeniably, the majority of highly successful onion producers in the community were of the Baptist persuasion and their attitudes to work contrasted strongly with those of their fellow-*comunarios*. Several non-Baptist elders maintained that it was pointless to 'try to become rich' because their neighbours would become jealous and they would be expected to sponsor a *fiesta* - the Aymara way of restoring the balance between rich and poor. Whilst community elders had traditionally welcomed the opportunity to improve their prestige within the area by acting as *prestes* (sponsors) for *fiestas*, it is certainly true that their families' food and financial resources could easily be whittled away. For example, one Llamacacheño *preste* in 1971 was obliged to make the following contributions: food (five sheep, potatoes, *chuño*, broad beans, maize, bread) and drink (five crates of beer, tins of alcohol and *chicha*). Additionally he had to pay for a band,

159

buy invitation cards, vestments for Compi's statue of St. Peter and sticks of dynamite for use as fireworks.

Not surprisingly, deeply engrained beliefs, together with a traditional fatalistic approach to matters of life and death, accounted for a number of seemingly-avoidable problems and disasters. For example, the conclusions of three UN veterinary surgeons, after examining sheep in Chua Visalaya, that an unusual number of deaths were largely attributable to poor husbandry and under-nourishment, rather than disease, occasioned no relief in the community: on the contrary, several stock owners remarked that such 'gross allegations' were 'slanderous'. Whilst they were more than prepared to accept that a misdemeanour of some *campesino* or family had evoked a curse on the community's sheep, they were indignant that any outsider should accuse them of mismanagement. One aged *ex-colono* was convinced that the sheep deaths were a punishment for numerous incidents of sheep theft : 'Once we've found out who's been stealing our sheep, we'll go and ask the *brujo* to put a death spell on him'.

In 1971, 18 Llamacacheño householders emphasized the need for a more reliable water supply: according to them, a wholly inadequate supply of water both for domestic and irrigation purposes acted as a veritable stumbling block to community development; in periods of drought, water supplies became critical. Yet several admitted the community had refused to collaborate with Peace Corps volunteers proposing to erect a water pump; 'some men didn't think it was necessary and weren't prepared to give any money'. Additionally, the NCDP had been willing to meet half the costs towards providing a sheep dip for Llamacachi but initial enthusiasm had waned once the TDC started approaching individuals about making financial contributions and participating in the building programme.

Countless examples of inequalities between the sexes in relation to farming were encountered in the lakeside region e.g., *campesinas* were usually paid half as much as *campesinos* (i.e. children's wages) for casual field work; women were considered 'too clumsy' to use any type of machinery; *campesinas* were frequently seen carrying extremely heavy loads of field produce whilst their husbands or teenage sons accompanied them on bicycles, and few girls were allowed to attend secondary school because it was 'a waste of time and money' and, in any case, they were required to help with field tasks, assist their grandmothers with herding and look after younger children. Women prepared to talk about their own situations identified a variety of daily problems: the difficulty of feeding the family during *la miseria* (the name given to the period of hardship and food shortage from November to April); the problem of looking after young children, especially carrying a baby on the back whilst weeding or clod breaking; backache from field work and carrying burdensome loads (including buckets of water) and the lack of electricity and labour-saving devices in the home.

Despite these and other difficulties arising from the women's 'double day', it was clearly apparent that in communities such as Llamacachi, Compi and Huatajata, the status of women was beginning to change radically and rapidly as a direct result of their marketing activities. As Clark (1970, p.72) observed: 'in terms of increased market participation, the peasant women have experienced a greater relative change than the men as a result of the land reform'. Against their better judgement, Llamacacheños were being forced into sending their daughters to secondary school to enable them to learn to converse fluently in Spanish and master the rudiments of accountancy - both considered essential for successful negotiating in La Paz.

Some Chuan *ex-colonos* berated the community's young men for being 'too lazy ...they never had to work under the harsh conditions of the *hacienda* ... they don't know what hard work is like'. Simultaneously, young, 'progressive' *campesinos* were criticized for trying to persuade their elders to conduct crop experiments when the shortage of land made it 'too dangerous to take the risk'. Whilst elderly *campesinos*, including the secretary general of the Chua cooperative, accepted their inability to communicate with officialdom in Spanish and to understand legal documents as a considerable handicap, they resented to an even greater extent the fact that some grandchildren brought up in La Paz had not been taught Aymara and were unable to hold even the simplest conversation with their lakeside grandparents. 'How can they be expected to understand the way we live and farm the land?'

Reasons for such different responses to agrarian reform

The most intriguing problem confronting the researcher during field investigations in 1971 was how to account for the marked differences in land usage and marketing activities between neighbouring communities displaying similarities in terms of population size and cultivable land resources. Why had Visalayans failed so miserably to avail themselves of improved transport facilities and rapidly expanding marketing opportunities, whilst Llamacacheños had taken full advantage of the changes produced by the process of agrarian reform and established a flourishing cash crop economy based on onions? Perhaps even more remarkable is the fact that a number of the differences observed in 1971 are still visible almost 50 years after the Agrarian Reform Law supposedly dealt a death blow to the *latifundio* system.

Whereas the vast majority of peasant syndicate members in lakeside communities had assumed ownership of nothing more than small patches of *ex-hacienda* land i.e. apart from household plots previously cultivated on a usufruct basis, Visalayans were considered by *campesinos* of other *ex-haciendas* in the area to have reaped a number of benefits from the break-up of the estate.

By his rare example of intensive and prosperous farming, the last *patrón* had laid the foundations for radical modifications in traditional peasant cultivation. Furthermore, on his untimely death in 1959, the community had gained access to an unknown, but large, number of pedigree cattle and sheep and an almost unparalleled collection of valuable agricultural machinery and timber cutting equipment.

No amount of questioning provided an entirely satisfactory answer although many relevant points were raised by the farmers of both communities. Unlike Visalaya, Llamacachi had a long tradition of exchange and marketing, enabling onion producers to build on earlier experiences. Whilst it was acknowledged that religious differences had created internal conflict, Baptist farmers were convinced that Llamacachi's connections with the Baptist mission in Huatajata had been extremely beneficial economically; in pre-reform times, Huatajatan Baptist missionaries had been active in persuading church members not to spend their hard-earned cash on festal drinking bouts but rather to invest it in farming improvements and durable goods. More than a few *comunarios* proudly boasted that the determination, stamina and resilience displayed by their ancestors in resisting the encroaching advances of neighbouring *hacendados*, provided ample proof of Llamacacheños' 'natural superiority'.

Most Visalayan *campesinos* were convinced that the intense rivalry between cooperative and syndicate members and the general lack of community cohesion, impinging on almost every facet of socio-economic life, were the main reasons for Visalaya's lagging behind Llamacachi and *ex-haciendas* such as Compi and Huatajata. Any profits made by the cooperative were still being absorbed by burdensome debt payments to the Agricultural Bank. In *hacienda* times they had been discouraged by the *patrón* from attending school; consequently, even the leaders of the cooperative and syndicate were unable to communicate directly with bank and agrarian reform officialdom. Most *campesinos* denied butchering estate animals for consumption at *fiestas:* they had 'died naturally'. They also insisted that the farm machinery had been allowed to rot because they lacked the training to maintain it though it was evident that most pieces of equipment had been 'cannibalized', presumably for sale purposes. It was claimed that the shortage of land prevented both syndicate and cooperative members from taking the risk of experimenting with 'new crops'; in any case, how could they afford to pay for seeds or fertilizers?

Lane Vanderslice, also carrying out investigations in the Lake Titicaca region in 1971, reached the conclusion that the *campesinos'* reluctance to adopt innovations in farming constituted Bolivia's most acute problem, since industrial growth could only be sustained by increased agricultural output. In a lecture entitled 'How to modernize the *campesino*' he identified the factors determining the level of acceptance of change amongst Jank'o Amaya's 300 families as follows: education, size of holding, age of household head, amount of contact with

La Paz and the impact of individuals, notably friends, neighbours and leaders of community opinion.

Unquestionably, though not mentioned as a significant factor by *campesinos*, the presence in the community of widely respected, innovative leaders, capable of acting as 'culture brokers', gave Llamacachi a substantial advantage over Visalaya and most other communities in the neighbourhood. Whilst Chua Visalaya's syndicate, partly by reason of its stormy career, failed to provide 'a good channel for brokerage', three individuals, drawn from Llamacacheño families returning to the lakeside after protracted periods in the city, played crucial roles in the development of the community's highly successful cash crop economy. Whilst the advice and practical assistance of the resident Belén-trained TDC had been readily sought and some suggested innovations (such as experimentation with improved seeds and fertilizers and the purchase of pedigree stock) put into effect, the writer's field assistant, educated in La Paz, had done much to raise the status of the community's women, especially onion vendors. She had acted as an intermediary in local disputes, pressed claims against fraudulent traders and truck drivers (during the writer's stay in the lakeside, she successfully brought a case against a driver for spilling kerosene over a sack of sugar purchased for sale in her parents' general store), been a spokeswoman for the community's onion sellers in the syndicate of the city's Avenida Montes market and negotiated favourable terms for lakeside stall holders: later she was to be elected as president of a 400 members-strong women's meat marketing syndicate in La Paz.

The impact of a community elder, one of the most progressive and efficient farmers in the vicinity, had been even more profound. He played a leading role in the rapid development of the onion trade, persuading fellow-*comunarios* that, despite their respect for *Pachamama*, it was essential to review ancient crop rotation systems. With the aid of the Baptist mission, he had opened the community's first school. His formal education in La Paz, bilingualism and intimate knowledge of judicial matters enabled him to offer legal advice within the region and equipped him for political leadership; he became the natural choice for Jank'o Amaya's first *intendente* (head of *cantón*). The situation in Chua compared very unfavourably.

Chua's former derelict *hacienda* house still dominated the lake shore in the early 1970s. By *ex-colonos* it was viewed as a constant, unavoidable reminder of the injustices of the feudalistic *colonato* system they had escaped from as a result of agrarian reform. More pernicious than any visible legacy of the *hacienda* system (though not recognized as such by the *ex-colonos* themselves) was the lasting impact of its excessive paternalism on *campesino* attitudes and behaviour patterns. Under the *colonato* system, to the *colonos'* detriment, they had not been required to make any responsible decisions: they had carried out the orders of the *patrón* or his staff to the meanest detail. However unpleasant living and working conditions might have been, successive *hacendados* had furnished most of their basic needs and afforded protection from the outside world. They had been unable to improve their work prospects through education and any tenant overtly demonstrating qualities of leadership and organization had been quickly removed from the scene or his activities curbed in other ways.

García (1970, p. 312) specified as one of the positive outcomes of agrarian reform 'the re-establishment of the confidence of the rural communities in their capacity to take the initiative and in the social importance of communal values'. Whilst a capacity or a willingness, to take the initiative might have been characteristic of *ex-colonos* in some communities, sadly the reverse was true for the most part in Visalaya, and numerous other communities in the lakeside region and on the Altiplano. For this situation, confounding the well-intentioned efforts of countless individuals and NGOs, the excessive paternalism of past *hacendados* was doubtless largely responsible. In the 1970s, and still in the 1980s, any outsider visiting Visalaya was regarded as a substitute for the *patrón* and accordingly expected to solve pressing problems and satisfy basic needs. The sentiment prevailed within the community, as in many others, that 'they' (the central government) should participate to a far greater extent in problem solving and providing various forms of rural infrastructure. The majority of elderly *campesinos* refused to believe that by their own endeavours they could improve agriculture or create a better life.

The Declaration of Tiwanaku

The *Manifesto of the Quechua and Aymara Indians*, signed at Tiwanaku in July 1973, and generally known as *The Declaration of Tiwanaku*, criticized successive Bolivian governments in much harsher terms. Whilst acknowledging that 'Agrarian Reform enabled us Indians to free ourselves from the ominous yoke of the *patrón*', it proclaimed that all Bolivian Indians continued to 'feel economically exploited and culturally and politically oppressed'. 'Social justice' in the full sense of the term remained elusive. The individual *campesino* was 'in all respects, a loser' - 'the veritable outcast of our society' and the victim of imposed 'pseudo development'.

> We peasants ... are foreigners in our own land ... We work simply to keep alive, and even this is often more than we can do ... The agrarian policy of our governments has been abominable ... We are abandoned to our own fate. Our country spends more than 20 million dollars to import agricultural produce. They prefer to pay money to people overseas rather than to their own peasants. When bank credits have been provided for the rural sector, they have helped only the new landowners and the cotton,

sugar cane and cattle barons ... it is the peasant who always comes out the loser, since he is the weakest (Materne, 1980).

Lakeside dwellers invited in 1981 to comment on the statements made by Indian leaders in the manifesto responded predictably i.e. in accordance with their family circumstances. Economically successful and relatively prosperous Llamacacheños and Compeños considered the viewpoints exaggerated. Clark's remarks (1970, p.71) reflected the reality of the situation for them: 'In terms of material comforts the northern highland peasant is much better off than previously ... the total value of goods purchased for consumption on a regular basis in markets for a family of five is ... three times more than the pre-1952 value'. On the other hand, poorer *ex-colonos* endorsed the Indian leaders' sentiments about the ongoing suffering and exploitation of the republic's indigenous peoples. All *campesinos* were adamant that the general lack of small-scale credit facilities available to farming communities remained a major problem.

'Material comforts'

During the 1970s Clark's 'material comforts' multiplied dramatically in the more favoured lakeside communities. After a visit to the study region in 1981, the writer was able to report (1983) thus:

> The recent changes along Lake Titicaca's shores are remarkable and during the last 10 years some communities have altered almost beyond recognition ... Today's visitor can not fail to be impressed by the veritable boom in house building: virtually every house near the lake appears to be in the process of acquiring a second, even third, storey, built of cheap red bricks rather than the traditional *adobe* ... Throughout the area external walls have been painted in vivid colours; curtains, spouts and gutters are no longer the exception. Several households have even created small gardens in which hollyhocks grow alongside the *kantuta*, ... Young married couples formerly living with parents are now constructing their own houses: consequently, the actual number of dwellings has more than doubled in some communities such as Huatajata. ... Electricity has also reached the lake. And with the electric light - enabling *campesinos* to go to bed later - has come the television. More than 30 families in Huatajata and 12 in Llamacachi have acquired black and white sets. Ovens run on liquid gas are now supplanting kerosene stoves and electric irons and fridges (bought cheaply on the black market in La Paz) have recently appeared on the scene. In the majority of roadside houses crude wooden

beds and mattresses made from flour sacks stuffed with straw have replaced *totora* reeds piled on often-damp *adobe* floors.

It is tempting to assume that such highly-prized acquisitions had been procured with the cash derived from farming activities. Whilst some undoubtedly had, most of the televisions, bicycles, fridges and stoves had been purchased either by city-based migrants in salaried employment or with the remittances sent by them. A number of *campesinos* acknowledged that they had come to rely heavily on money and consumer goods supplied by migrant children. At the same time, some of the more successful onion vendors of 1971, capitalizing on their entrepreneurial skills and experience, were by the early 1980s operating their own businesses e.g. selling liquid gas, running general stores or selling woven artefacts direct to tourist shops or stalls. Near Huatajata a returned migrant woman had even opened a bakery employing four men and serving the lakeside communities between Huarina and Tiquina. Several such women confessed to hoarding some of their earnings to finance their daughters' education. As one commented: 'Why should my husband spend my hard-earned cash on drink?'

10 Some conclusions to be drawn from the case study

Sustainable agriculture

True progress must be based on a culture. It is a people's most basic value. The root of our national frustration has been that Quechua and Aymara cultures have always been the object of a systematic drive to destroy them. The politicians ... have sought to promote a development based wholly on a slavish imitation of other countries' styles of development, even though our cultural heritage is quite different ... We peasants desire economic development, but it must be rooted in our system of values. We do not want to lose our noble ancestral virtues on the road to a pseudo-development..We must improve on our past with technology and modernization but in no way break with the past (The Declaration of Tiwanaku, 1973).

Today's economists generally attribute 'the backwardness' of *campesino* farming in highland Bolivia to 'primitive technology', the lack of mechanization, low levels of investment and an inadequate infrastructure. In the 1997-98 'Country Profile of Bolivia' published by the UK's Economist Intelligence Unit, the subsistence farming of 'the primitive traditional highland areas and central valleys' is sharply contrasted with 'commercial agriculture' which has 'in recent years taken strong root in the eastern lowland department of Santa Cruz'. Indeed, by 1997 soya had become 'Bolivia's most valuable export, ahead of both zinc and gold'. Whereas 'the use of primitive technology in subsistence farming' is blamed, together with illegal deforestation, for 'causing serious environmental degradation', there is no reference to the long-term adverse effects of monocultural practices, the excessive application of chemicals and the potentially harmful environmental consequences of introducing genetically modified soya to the eastern plains.

Whilst the economic success of commercial farming (with strong financial support from the United States, Brazil and Argentina) is indisputable, the dismissal of traditional Aymara and Quechua technology as 'primitive' is wholly unjustified. The overwhelmingly dominating conclusion to be drawn from this survey of agricultural change in the Lake Titicaca region is that few, if any, of the modifications to traditional farming practices, introduced in the name of agrarian reform and agricultural development and still visible at the end of the twentieth century, compare favourably with the technological achievements of, and sophisticated techniques practised by, Tiwanaku, Aymara and Inca/Quechua peoples at least 450 years ago. Virtually everything that has happened in farming since the Spanish Conquest is more appropriately termed, agricultural change than agricultural development. The more details archaeologists unearth about pre-Conquest farming, the more researchers can only marvel at past accomplishments. As Luis Lumbreras, a Peruvian archaeologist, commented prior to the Columbian Quincentenary: 'we are still blind to the misdirection of our development. The Andean world remains impoverished because we are unable to see except through colonial lenses' (1991, p. 22).

In a 1985 article entitled, 'Peru's Agricultural Legacy: Ancient methods may be useful in reviving today's food production', William Denevan drew attention to the fact that: 'Areas of prehistoric cultivation, requiring the movement of vast amounts of earth and stone, dwarf the better known pyramids and cities of the ancient New World'. Whilst pre-Columbian farmers may be accused of re-modelling, if not re-constructing, Andean landscapes, it should be remembered that they did so in order to practise elaborate forms of sustainable agriculture, adopting precautionary measures for preserving soil fertility (e.g. by strictly observing fallow periods and rotation systems) and conserving water, flora and fauna resources. Over the centuries they had devised and perfected techniques for exploiting fragile environments sustainably, whilst simultaneously overcoming seemingly insuperable difficulties imposed by them. Irrigation channels, the construction of which entailed considerable engineering feats, extended areas of cultivation, whilst, as Lumbreras notes:

The *"andenes"*[the walled terraces so widespread in the study region] represented a productive strategy for maximum utilization of the scarce water resources of the central Andes. They made it possible to prepare lands on the slopes for sowing without serious dangers of erosion ... In the Andes such terracing was a momentous discovery, which our Western mentality has yet to appreciate ... the West did not know what to do with the terraces and classified them as "primitive".

At four locations near the lakeside communities, impressive research is currently being conducted by the NGO, PROSUKO (Inter-Institutional Program of *Suka*

Kollus), on the feasibility of reviving Tiwanaku-era raised field systems, known locally by Aymara farmers as *suka kollus* (in Quechua as *waru waru*, and in Spanish as *camellones*). Aerial photography has enabled archaeologists to identify one area of *suka kollus* covering some 60,000 hectares of land near the lake-shore. Whilst the national average yield of potatoes per hectare is 3.7 tons, one community participating in recent experimental work obtained an average yield of 34 tons per hectare. The 'sophisticated hydraulic system' producing 'the abundant harvests, maintaining the people of Tiwanaku for more than 1,000 years' was described thus by CEDOIN in a 1989 bulletin on water management:

> The raised fields, some as large as 50 feet wide and 600 feet long [15.2m x 182.9m], are layers of stone, clay, gravel and topsoil to a height of three to five feet .... irrigation channels running between the fields stored water for use during the dry season and drought periods. In addition the sun-heated channel water served the important purpose of protecting crops from frost damage during the bitter altiplano nights. Algae and other aquatic plants grew in the channels providing food for fish, which in turn encouraged a duck population. Decayed plant and animal material formed a thick sludge which not only helped to store solar heat, but was also used as a fertilizer for the fields.

Although it would be possible to apply the ancient well-tried techniques in suitable locations round the lake, it remains to be seen whether *campesino* communities maintain the field systems once PROSUKO has withdrawn financial and advisory support. *Suka kollus* unless they are constructed on communally-held land, require 'pooling' land resources, which has proved unpopular in the Lake Titicaca region since 1953; preparation and maintenance also depend on a heavy labour input, not readily available in most communities.

Pre-Columbian farming communities had also developed unique techniques for dehydrating tubers, by taking advantage of the extremely wide diurnal temperature range in mid-winter; the same methods used today enable *chuño*, *tunta* and *c'aya* to be stored in confined spaces for several years without rotting - a major reason for food shortages elsewhere in Bolivia, as throughout the Third World. Andean farmers had selected highly nutritious plants to cultivate for human consumption, were using a large number of herbs medicinally and had discovered plants that could be used highly effectively as pesticides and insecticides: in some valley communities several varieties of mint are still burnt near certain crops to ward off potentially damaging insects,

Precisely what size of population ancient farming systems in the lakeside region supported in pre-Conquest times is impossible to establish with any degree of accuracy. The work force required, even over a number of years, to construct the walled terraces illustrated in the aerial photograph suggests that pre-Columbian populations were considerably larger than at present. Recent research has led archaeologists to conclude that Pampa Koani, by the lake, 'where about 7,000 poor Aymara farmers live today, produced enough to feed and support over 125,000 people in Tiwanaku times' (CEDOIN, 1989).

With the arrival of the Spaniards, 'ingenious native irrigation systems and immense areas of cultivation were abandoned to sow European plants' (Lumbreras, 1991). In *hacienda* times examples of highly productive farming were rare on the Altiplano. The only one encountered in the study region was *Hacienda* Chua, during the ownership of the last *patrón* - the *hacienda* described in earlier chapters and known to SNRA personnel for its successes with livestock rearing. Yet it has to be acknowledged that high production levels were bought at a price i.e. in addition to unpaid, and otherwise exploited labour. A number of the practices adopted by the last landlord did **not** promote sustainable farming or sustainable management of the environment in the long term. Large flocks of sheep contributed to the degradation of pasture lands, whilst the introduction of tractors set in motion today's acute problems of topsoil loss by exposing sizeable areas of deeply ploughed land to the damaging effects of winds and storms. Furthermore, the *hacienda* was one of several in the region to popularize the establishment of eucalyptus plantations. Eucalyptus trees were planted for aesthetic reasons and to create wind breaks; they provided a useful resource of fast-grown timber, but were to prove totally inappropriate in terms of their water requirements (up to 500 litres per tree per day) in an area prone to drought, where supplies of fresh water are normally at a premium.

Communications between *campesinos* and outside agents

Between the early 1960s and mid-1980s criticisms of Bolivia's agrarian reform policies tended to focus on a wide range of issues viz. the complex and lengthy procedures involved in land expropriation and redistribution, colonization projects, the failure of peasant cooperatives, the inadequate provision of rural credit facilities and extension services, and statistical data relating to agricultural productivity: by the late 1980s, excessive bureaucracy, institutional corruption, *minifundismo, neolatifundismo* and the need for modifying the original agrarian reform law dominated analytical research and discussion. Over the years the study region has yielded abundant evidence in support of the writer's conviction that one of the crucial, yet frequently ignored, factors determining the eventual outcome of even the most carefully planned agricultural project is the intangible one of human interactions. Since 1953 conflict within peasant communities and misunderstandings between farmers and outside agents (government personnel, extension agents and NGO employees) have constituted a major obstacle to both agricultural and community development in many parts of rural Bolivia - as throughout Latin America.

170

According to Conyers (1982, p.130), 'the gap' between participants in the change process and outsiders 'leads not only to practical difficulties in communication, such as language problems, but also to differences in attitudes and expectations and to mutual feelings of suspicion, mistrust, resentment and even derision'. In the early days of agrarian reform, 'the cultural gap' was extremely wide between *campesinos* and urban based agrarian reform officials, many of them with no experience of farming and rural life, and unfamiliar with Aymara and Quechua thought processes in relation to land and the supernatural world. A number of problems could have been anticipated, and some failures prevented, by making early contact and establishing effective communication links amongst the various interested parties. As several lakeside *campesinos* remarked in 1971, they could not expect government officials to understand the nature of their age-old attachment to the land and their attitudes to farming practices, but had *colonos* and *comunarios* been asked to air their opinions in 1953 about migrating to colonization zones in either the Yungas or eastern lowlands, they would have had no hesitation in stating categorically that the vast majority of lakesiders were not the least interested (for reasons given on p.144) in taking up government offers of 'free land' in new settlement areas.

It is clear that had the Commissioners drafting the 1953 Agrarian Reform Decree been better informed about the average size of communities and land holdings in different parts of the republic, they could have arrived at more realistic, less controversial proposals for the maximum extent of the various categories of holdings. Whereas *campesinos* in the Lake Titicaca region were told they were entitled to 10 hectares of land, land scarcity made it impossible for countless families to obtain titles to more than two hectares of land. It is no exaggeration to say that disputes over entitlement to land permeated virtually all aspects of life in this part of Bolivia, as in other densely populated regions, throughout the 1960s and 1970s. As already noted, much time and energy was expended on brawls and reunions at which 'the land question' was debated endlessly. The government's gross miscalculations and unfulfilled promises must be held largely responsible for the degree of discord which frustrated many worthwhile development schemes and projects supported by government representatives and NGOs.

During the 1960s and 1970s the Ministry of Peasant Affairs and the NCDP set great store by the extension agent (TDC): persuasive, articulate *campesinos*, selected from and respected within their areas of operation were expected to function as 'catalysts of modernization'. The TDC working in the late 1960s in the Chua-Llamacachi area assisted a number of the wealthier, more progressive farmers with the purchase of improved seeds, fertilizers, pedigree pigs and sheep (p.146). Yet, the very fact that he was a government employee living in 'comfortable' circumstances alienated him from virtually landless *campesinos*. They considered such an agent to be 'necessary' but complained that they had no cash to take advantage of his services i.e. that they had no alternative to planting already infected seeds and rearing poor stock. For his part, the TDC admitted to feeling extremely frustrated by the *campesinos'* unwillingness to take advice and collaborate, especially on projects designed to satisfy felt needs. He alleged that the main problems in his extension work were posed by the *campesinos'* traditional fatalism and superstition - no less formidable problems today.

Campesino cooperatives and syndicates

According to the Declaration of Tiwanaku (1973), 'the cooperative system is innately natural for a people which originated mutual-assistance means of production ... Private property, political partisanship, individualism, class distinctions, and internal struggles were introduced to us by the colonizers and enhanced during the republican period'.

Twenty years earlier Paz Estenssoro and the MNR had expressed their confidence in, and commitment to, agrarian cooperation, convinced it had the ability to raise agricultural productivity (p.63). The cooperative had been seen as the appropriate, if not ideal, labour unit for conducting crop experiments and introducing new farming techniques and technology. Despite such enthusiasm, agrarian cooperatives in Bolivia, as in a number of other Latin American republics, were to prove a great disappointment to their advocates; by the early 1970s the majority of them were defunct.

Over the years a number of researchers have maintained that the cooperatives formed as a result of agrarian reform were doomed from the start by a combination of factors. Some have doubted the validity of the premise on which they were based - a premise nevertheless upheld in the Tiwanaku manifesto. When challenged on this point in 1981, lakeside *campesinos* affirmed that traditional cooperative farming in the Andean regions is a 'myth' - their ancestors had no choice in the matter. Work was highly 'regimented' and they were **forced** to work collectively by *kurakas* or Inca chiefs and later by *hacendados*. They had always worked 'naturally' in extended family units, with men, women and children assigned specific tasks - not as groups of men 'having to do women's work'.

García (1970, p. 331) blamed the peasant syndicates for the failure of Bolivian cooperatives: the 'subordination' of the cooperatives to rural syndicates 'not only diverted them from their own ends, but also stifled their autonomy and closed the road to a new entrepreneurial economy'. The syndicates themselves, responsible for initiating *dotación* proceedings, 'were rooted in the struggle for land and peasants' rights'. As elsewhere in Latin America, Bolivian syndicates were essentially political and not 'created to serve economic functions such as joint efforts directed at obtaining credit, marketing products, or purchasing inputs and supplies' (Dorner, 1992, p.51). Bebbington (1998, p.175) criticizes them further for paying 'little attention to natural resource management, technology development, or issues of income generation'.

The study of Chua Visalaya has illustrated the difficulties and frustrations arising from co-existence: tensions over membership, landownership and community decision making were predictable. It was almost inevitable that cooperatives and syndicates would become rivals. Comments made by *campesinos* confirmed Paz Estenssoro's views about the importance attached by *campesinos* to individual ownership of land. Whilst Aymara farmers retain their feelings of respect and reverence towards *Pachamama*, attitudes towards title rights over plots of land have changed dramatically over the centuries. An OXFAM UK project initiated in 1971 'to investigate the relevance of cooperatives and possibilities they offer for solving some of the socio-economic problems of the rural people' of La Paz department and, additionally, to organize elementary accountancy courses for cooperative leaders, established that at least 22 of the cooperatives officially listed no longer existed. They had been dissolved as a result of misappropriation of funds, lack of confidence in leaders, fears of 'being tricked and exploited' and a general unwillingness to plough back early profits for long-term investment. Dorner (1992, p.54) identifies 'the lack of accountability of management' as one of the key problems with production cooperatives. Commenting on 'the breakup of cooperatives' in a number of Latin American countries, he stresses that whilst production cooperatives have experienced their 'share of problems', other forms of 'cooperative efforts among and between farmers are vital'.

Rural education

Not surprisingly, the Declaration of Tiwanaku roundly condemned Bolivia's rural education system, at a time when it was obligatory to conduct all teaching in Spanish:

> It is no secret to anyone that the rural school system is not based on our cultural values ... Rural education is a new and more subtle form of domination...the teaching is devoid of roots. It is unrelated to our life, not only in its language but also in the history, the heroes, the ideals and the values it presents.

In common with farming communities near cities in many parts of the Third World, lakeside *campesino* communities have over the years adopted an ambivalent approach to rural education, as to improved transport facilities. Whilst secondary education continues to be regarded as vital to social and economic mobility, simultaneously it has been perceived as posing a major threat to future farming prospects. Undoubtedly increased access to education is in large measure responsible for the marked decrease in the agricultural labour force. Not only has it stimulated rural-urban migration by raising expectations and aspirations, but it has also been instrumental in reducing average family size. No adult interviewed in the study region in 1981 wanted more than two or three children - a far cry from the situation a decade earlier, when one Visalayan *campesino* had hoped to father 20 children in order to 'prove' his 'virility to other *campesinos*'!

Whereas since the 1960s the processes of 'conscientization' and 'popular education' (both encouraging respect for, and preservation of, indigenous cultures and languages) have been used by NGOs in both rural and urban communities to empower grassroots groups, since pre-reform times fluency in Spanish has been considered both by cash crop producing farmers and the younger generation as a highly desirable acquisition. As previously noted, a working knowledge of Spanish was by the 1970s deemed essential for successful city marketing and urban employment generally. By 1981 teenage boys were intent on obtaining 'good grades' to enable them to enter the police force or teaching profession. Several girls from the lakeside communities had by then trained as policewomen and increasing numbers were entering rural colleges to train as teachers, though none of those encountered intended to apply for teaching posts in the countryside. In 1995, largely at the instigation of Bolivia's Aymara vice president, elementary teaching in indigenous languages was legalized. Whilst in the more isolated rural areas communities appeared to welcome the change, many urban-living teachers working in lakeside schools and young parents, themselves educated in Spanish, regard the introduction of bilingual teaching in some ways as a retrogressive step - as a subtle means of marginalizing rural children.

Since the withdrawal of government extension services in the mid-1980s, the communication gap between *campesinos* and agrarian reform institutions has widened. One of the most pressing needs in farming communities today is for extra-mural courses for *campesinos* on farming issues having a direct bearing on their everyday lives e.g. means of conserving the soil, animal husbandry, prevention of crop and animal diseases, appropriate technology, water management, accountancy and ways of responding to changing monetary and marketing situations (discussed at the end of the chapter). All farmers interviewed in the 1990s share these concerns and would welcome information and advice. Whilst the more fortunate communities receive help from NGOs or other outside agents, the majority do not and find the costs of securing expert advice (e.g. from vets) prohibitive. Although at the time of the writer's visit in 1998, the ALT barrages were being constructed across the River Desaguadero, even the few *campesinos* from the study region who had seen the work in progress had no idea of its purpose, despite the fact that it could have a significant bearing on their future livelihoods.

Extension services

Since *Ley INRA* has failed to respond to *campesino* demands for technical assistance, farming communities throughout the republic are obliged to rely on the advice and support of NGOs and university agronomy departments. In *The National Directory for NGOs in Bolivia (1997)*, being consulted by government departments in 1998, five national and two overseas NGOs (one of them CARE International) were listed as running projects or programmes in the municipality of Achacachi, Omasuyos. It has to be said that no evidence was found of any of the five NGOs supposedly involved in agricultural development working in the study region. It is apparent from the directory that the distribution of NGO projects is extremely 'patchy'; 300 NGOs are listed for La Paz department, but only 12 for Potosí and 3 for Pando.

Of the universities running agricultural research programmes and conducting extension work in Achacachi municipality, the most active is St. Andrew's University, La Paz (UMSA). Since 1984 the university's faculty of agriculture has managed the Belén research station near Achacachi. Whilst staff members are primarily concerned with promoting agricultural development in local communities, they also supervise three research projects on stock and fodder, agroforestry and Andean crops. Additionally, researchers are creating a *banco de germoplasma*, with an emphasis on native food plants, especially *papa, papalisa, quinua, cañahua, oca, isañu* and *tarwi*. At the time of writing, experiments are underway to produce higher-yielding strains of *quinua* that are also resistant to mildew. The growth of *quinua* on currently disused terraces away from the lake, for marketing commercially, **could** provide a welcome source of income for struggling lakeside communities, at a time of tremendous international interest in the plant's nutritional value. At present 95 per cent of exported *quinua* (mainly grown in Potosí department and marketed by ANAPQUI i.e. the National Association of Quinua Producers) is destined for the United States; of recent years Germany, Austria, Belgium, Holland, France and Japan have also entered the market.

One of the encouraging aspects of agricultural extension work on the Altiplano at the present time is the fact that UMSA, The Catholic University of La Paz and several other universities with agronomy departments have in recent years established research centres in or near small rural towns (even at Tiwanaku), where agronomy students drawn from rural communities are based for periods of varying length . Whilst staying at the centres, they are able to carry out research for dissertation purposes and have the opportunity of working alongside local *campesinos* on agricultural projects. Students encountered in 1998 at three such centres (including one at Chulumani in the Yungas) remarked that it was their intention on graduating to return to their communities and put some of their newly-acquired knowledge and skills into practice. Whether or not they do so is likely to depend to a large extent on the land and financial resources available to them. It is to be hoped that all such graduates in practical agronomy are given every incentive to stay on the land: because of excessive rural-urban migration, Bolivia is desperately short of knowledgeable, enthusiastic, progressively minded young farmers, with the capacity to act as leaders and innovators.

Community development

Undoubtedly one of the main reasons for the withdrawal of NGOs from lakeside communities over past years has been the high incidence of conflict inside communities and the reluctance of the *campesino* population to work productively together on community development projects. Community cohesion, shared goals and community pride are essential to community welfare and development. 'Implicit in the theory that has been built up in relation to community development is an organic and physical concept of community - a group of face-to-face contact, bound by common interest and aspirations' (United Nations, 1971): in such an ideal situation the community developer's task would be straightforward. In the real world perfect harmony is a rarity. Conyers (1982) is convinced that: 'in all communities there are individuals and interest groups with different, often competing or conflicting, aspirations and objectives'. Other researchers have alleged that some measure of conflict and competition is necessary to stimulate change. Whilst both of these claims may be true, there is no denying that antagonism and physical confrontation in the lakeside have proved detrimental in many respects: NGOs with limited personnel and financial resources are not prepared to waste their efforts and time on communities where householders refuse to collaborate on routine tasks or to make minimal financial contributions within their means.

Likewise, avoidable mistakes and miscalculations have been made on countless occasions, often by well-meaning outside agents: consequently all such contacts are now accompanied by suspicion and mistrust on the part of *campesinos* - a practical problem ignored in agricultural extension work at considerable cost. In some situations individuals feeling 'threatened' in some way by outsiders can rapidly instigate antagonism. In 1981, after seven deaths from mumps in one community, the researcher enlisted the aid of the doctor responsible for the care of *campesinos* in the region: on the subsequent death of one of the children medically treated, the researcher was accused by local *curanderos* (curers) of 'making a pact with the doctor to kill all the children'.

At the same time it is evident that three factors have in some measure served to alleviate this counter-productive situation and encourage meaningful agricultural and community development. Firstly, it is a well established fact that innovation often occurs as the result of observing and following successful examples. Thus after the lakeside community of Chua Cocani had installed a water supply in 1981, once the obvious advantages, in terms of irrigation, hygiene and health care, were established, neighbouring communities decided to cooperate with CARE and the departmental development agency in making the installation of a potable water supply possible.

Secondly, rivalry between communities and community pride have on occasions cut across internal friction. When in 1971 Llamacacheños were asked why they wanted electricity, a frequent response was:'We want it because they haven't got it in Compi'. Various communities' pleasure in their schools' achievements, the building of a plaza and inter-community football matches have been instrumental in fostering, at least temporarily, a spirit of unity and strength of purpose.

Finally, much has been achieved through the determination of *campesinas*, some of them having learnt to act assertively against authority as members of city-based marketing syndicates. Before the 1980s lakeside women suffered individually in silence: their opinions were largely disregarded by elders and rarely sought by government officials. In several cases since the early 1980s when men have failed to reach unanimous decisions, for example on whether or not to install water supplies, groups of *campesinas* have decided to take matters into their own hands, pressurizing community elders to take positive action. In the case of water supplies, they have complained that it is the women who have to carry water for long distances and look after children suffering from water-related diseases; if the men refuse to do the work and contribute the money required, the women will do it themselves. In 1991 the writer attended a fascinating meeting held by a women's grassroots group in a *zona* of La Paz. The rural-urban migrant women present, some of them from the Lake Titicaca region, expressed their deep fears about the possibility of cholera gaining a hold in the city: eventually a unanimous decision was reached to force the *zona's* men to remove the large piles of disease-threatening rubbish from the neighbourhood streets by refusing to cook them any more meals until they complied with the group's request - a drastic measure which had equally dramatic results!

The Bolivian NGO, SEMTA (Multiple Services in Appropriate Technology), which has been operating for almost two decades and is currently working in 29 farming communities in Pacajes (one of the southern provinces of La Paz department) stresses the importance of '*caminando juntos*' (walking together) 'with *campesinos* and communities in their search for innovative ways to breathe life back into the Bolivian altiplano' (Campfens, *BT*, 1998). The organization is one of few in Bolivia advocating the application of appropriate technology and a wide range of farming activities together constituting agrarian reform and promoting sustainable agriculture: 'soil conservation and pasture restoration, preservation of native Andean crops, the improvement of farmers' agricultural and livestock production, commercialization, water collection, micro-credit schemes and training' (Campfens).

Farm incomes

It is no easier today than in 1971 to determine average farm incomes in individual communities with any measure of accuracy. Whilst *campesinos* are generally aware of how much money they have obtained in the current or past farming year from one-off large sales (e.g. of a cow) and can usually reckon their earnings from selling potatoes at a set price per *arroba* (11.5 kgs) or sack, very few keep any record from one year to the next. In the lakeside communities familiar to the writer only a handful of the poorest *campesinos* have over the years been readily prepared to divulge their farm income in the presence of strangers: a number have expressed their fears of unwanted repercussions - such as 'taxing by the government' - if they do so. The wealthiest *campesinos* have been anxious to withhold income details from their neighbours to avoid retribution - usually stealing or arson - resulting from their jealousies.

Difficulties in calculating incomes are further confused by bartering considerations and arrangements whereby food is given or acquired in exchange for some or other form of service. As previously noted (p.80), comparisons between past and present incomes can be meaningless in a country like Bolivia where prices have been known to fluctuate violently over a period of several years - or even months. Likewise, yields are guaranteed to vary dramatically as a result of climatic extremes, especially in a region subject to the vagaries of *El Niño*. Crop diseases and infestations, inadequate application of fertilizers, animal malnutrition and disease are all likely to seriously reduce farm incomes, although in situations where the problems are widespread higher market prices due to shortages may compensate for low production levels. Farm incomes in the Lake Titicaca region are also distorted by a variety of work practices e.g. the irregular participation of rural-urban migrants in field tasks, such as harvesting, different share-cropping systems, and lakeside dwellers' periods of temporary employment elsewhere, especially in La Paz, El Alto or the Yungas.

Globalization and the *campesino*

Bolivia's history of agricultural change and agrarian reform yields abundant evidence of the enduring intricate links between 'the land question' and national politics. Since the days of Bolívar, the farming prospects of the country's landowning and peasant sectors have remained largely at the mercy of successive governments. Whilst right wing, particularly military, regimes have in general lent support to the expansionist endeavours of the landholding elite, more liberally inclined governments, with a concern for social justice and welfare, have usually attempted to promote the rights and interests of the struggling peasantry.

Apart from political constraints, today's *campesinos* are confronted by almost insurmountable problems and threats to their livelihoods in the form of uncompromising, powerful market forces, the products of globalization. Bolivia's severe economic recession of the early 1980s, due in part to excessive government borrowing in the 1970s for funding prestigious projects, and exacerbated in the late 1980s by the tin 'crash' and collapse of manufacturing, dealt a death blow to many of the agrarian reform gains of the previous three decades. Small farmers were deprived of virtually all the inputs, such as government extension services, to which they had formerly been entitled and had access. No less damaging was the introduction of trade liberalization measures, paving the way for the import of cheap foodstuffs and food donations, both seriously undermining traditional market pricing mechanisms.

It would be logical to assume that the republic's expanding urban population guarantees rural producers - especially those farming in close proximity to large urban centres - a ready market for their commodities, providing them with an incentive to increase agricultural output. Bolivia's present national population growth rate of 2.6 per cent exceeds all others (with the exception of Paraguay) in South America, whilst its urban share has risen from 46 per cent of the total in 1985 to 58 per cent in 1998 (World Population Data Sheet). Instead, the austerity regulations (including the removal of food, transport and electricity subsidies) introduced in 1985 have cancelled out any gains in terms of potential market size, by severely restricting the purchasing power of all but the wealthiest urban dwellers. Consequently, a 1996 government report claimed that between 40 and 50 per cent of Bolivians were consuming less than 1,300 calories per day.

Investigations carried out by the writer in 1990 and 1991 in several *zonas* of La Paz and El Alto, in connection with an International Geographical Union research project on 'the impact of global economic restructuring on women's lives', revealed a number of food procuring survival strategies low-income urban women, especially household heads, had been forced to adopt i.e. in addition to paring household expenditure to the bone, chewing coca leaves to ward off hunger and withdrawing daughters from school to help boost family incomes by selling items on the street. In some cases rural-urban migrant women without direct access to family produce from the countryside were regularly visiting *tambos* on city outskirts in the early hours, to enable them to purchase food at the prices made available to intermediaries. Other were buying poor quality food (e.g. rotting fruit) at knock-down prices or collecting discarded foodstuffs (e.g. unsold meat after several days' exposure on dirty city pavements, or trichina-infected pork). Since the mid-1980s there had also been a phenomenal expansion in the cooking of low-cost meals, especially 'thin' soups, on the street. Not surprisingly, eating smaller quantities of food, ill-balanced diets, contaminated and diseased food, inadequately cooked meals (cases of gastro-enteritis and salmonella are frequently related to street cooking) have emphasized the problems of malnutrition and poor health, in a country with the lowest life expectation (60 years) and the highest infant mortality rate (75 per 1,000) in South America. Researchers, conducting a survey for the government's Sanitary Emergency Unit, concluded in March 1999 that 45 per cent of all food sold in la Paz is in some way contaminated.

In late September 1998 Bolivia was granted by the International Development Bank, the IMF and World Bank 'a total of US $760 million in debt service relief' under the Heavily Indebted Poor Countries (HIPC) initiative. According to the World Bank president, the decision marked 'a vote of confidence' in the Bolivian government's prudent economic management. The money 'saved ... is supposed to be invested in social programmes' (Andean Group Report, October 1998). It is to be hoped that some of it will be allocated for projects or programmes calculated to improve the well-being, or raise the incomes, of *campesinos*. In an article, 'Globalisation, Peasant Agriculture and Reconversion' (1997), Kay discusses the 'series of measures' introduced with the aim of 'enabling and improving peasant agriculture's ability to adapt to Chile's increasing exposure to global competition and to enter into the more dynamic world market'. A debate on these vital issues is long overdue in Bolivia.

11 The crisis in *campesino* farming in the Lake Titicaca region at the end of the twentieth century

Tourists travelling through Bolivia's Lake Titicaca region on a once-in-a-lifetime visit to South America marvel at 'some of the most dramatic and beautiful scenery in the world' (Cox and Kings, 1998). They are impressed not only by the splendour of the lake but by the 'idyllic' scenes of *campesino* families 'working cheerfully' in the fields, by the 'quaint houses', the bustling weekly markets full of produce and the colourful *fiestas*. The majority of lakeside communities are seen to contain brightly painted schools: near Huatajata a rural hospital (built by a Spanish NGO) opened at the end of 1998. A number of communities are in the process of constructing or up-grading their small *plazas*. Several impressive houses are being built along the shores of the lake. All of the communities skirting the through-road now have electricity and potable water supplies, whilst the road itself is well maintained and provides direct access to El Alto and La Paz.

Huatajata, with some 20 'restaurants' catering for the needs of pilgrims and tourists, demonstrates every sign of being a thriving tourist centre. Visitors are able to visit the *Museo de Balsas Trans-Oceánicas*, a reed boat museum run by the family of Paulino Esteban, one of the Suriqui islanders originally taken by Thor Heyerdahl to Morocco in 1970 to construct RA II. Since the lake floods of 1985-87 partially destroyed Guaqui, formerly Bolivia's lake port for passenger boats, Huatajata - endeavouring in the early 1970s to become a 'new town' - has emerged as Guaqui's successor: its five-star hotel organizes daily hydrofoil services to Copacabana and Puno, Peru's main lake port and, with a population of 80,000, the lake's largest settlement. Not surprisingly it is generally assumed by visitors to the region that lakeside dwellers must be more than content with their lot and that living conditions are continuing to improve with the passage of time.

To the environmentalist, the agrarian expert, the outsider familiar with the communities - and, not least, the lakeside dwellers themselves - the reality of the situation is very different. Whilst a few elderly *campesinos* (including a number of those able to recall with clarity their early years as *colonos*) persist in describing life as being 'like a flower', 'very pleasant' or even 'perfect', the majority refer to it as 'hard', 'painful', 'back-breaking', if not 'impossible'. Most are adamant that incomes from farming have declined dramatically and life has become considerably more difficult since the introduction in August 1985 of Bolivia's neoliberal New Economic Plan, especially the imposition of austerity measures. Some *campesinos* insist that their families survive on an annual income not exceeding US $ 500; Bolivia's average income per head in 1998 was US $ 830, the lowest in South America, with the exception of Guyana.

Visits to the area in 1990, 1991, 1996 and 1998 have left the writer feeling increasingly pessimistic about the *campesinos'* prospects for farming in the twenty-first century. It is clearly apparent that, after dominating life in this part of the Andean region for more than 3,000 years, agriculture has reached a critical point. Many centuries-old Aymara communities today face a bleak and uncertain future. The very survival of some is threatened by the expansion of tourism, by speculators, would-be commuters and second-home owners, competing in the land market for highly desirable building sites (presently cultivable plots), at a time when rural-urban migration has drawn away from the countryside its next generation of farmers. The title of this chapter is that of a seminar given in August 1998 at the Bolivian Academy of Science, in connection with a British Council academic link between Glasgow University geographers and the staff and students of UMSA and the Military School of Engineering (EMI), La Paz.

On that occasion the following major reasons for concern were identified and discussed: severe soil erosion and land degradation generally, in an area characterized by extreme *minifundismo*, where *campesinos* had no prospects of acquiring additional land through the process of *saneamiento*; land and water contamination affecting crop production and animal husbandry, in addition to human health; lack of a co-ordinated irrigation policy, abandonment of pre-colonial practices and misuse of limited water resources; scarcity of credit facilities and withdrawal of government support, including the provision of extension services, with disastrous implications for subsistence farmers; marketing problems e.g. pricing mechanisms, competition, food donations from overseas and growing preference for imported pastas; calamitous effects of excessive rural-urban migration on community and family farming; the inability of rural education to provide incentives for young people to stay on the land; the unprecedented growth of tourism, diverting land and financial resources from agricultural usage; increasing competition for land from speculators etc., loss of farm land for house building, road widening and as a result of gravel extraction;

the threat of land expropriation for environmental reasons and the ever-present threat to agricultural production posed by extreme weather events.

To place these problems - many of them by no means new but increasing in magnitude - in order of priority is extremely difficult. For example, although no *campesinos* interviewed in 1998 could envisage a lakeside without farming as its main occupation, now that *Ley INRA* has authorized individual *campesinos* to sell plots of land, albeit at fixed prices, land sales could become the crucial issue in the near future. On the other hand, it is conceivable that regulations governing the sale of land could be reversed by a future government.

El Estado del Medio Ambiente en Bolivia (The State of the Environment in Bolivia), published by the NGO, LIDEMA (Alliance for the Defence of the Environment) prior to the 1992 Rio conference, emphasizes the severity of the republic's soil erosion and land degradation problem:

> Soil erosion constitutes Bolivia's principal ecological problem, by reason of its characteristics, size and implications for the rural economy and the feeding of its population ... Accelerated erosion is turning some regions of the country into desert, the worst affected being the Altiplano and some agro-industrial zones of Santa Cruz.

In a similar vein ALT's 1995 strategic plan identifies erosion as 'one of the Altiplano's greatest problems and the most difficult to resolve', noting that around the margins of Lake Titicaca soil erosion has accelerated very significantly over recent decades with the result that 89 per cent of the land is now classified as moderately, severely or very severely eroded.

Whilst recognizing that human mismanagement of the land had its origins under Spanish colonial rule, LIDEMA places much of the blame for the increased rate of erosion on the 1953 Agrarian Reform Law: 'the excessive fragmentation of land has greatly contributed to the over-usage of soils, over-grazing and the abandonment of ecologically better adapted pre-colonial practices'. The ALT survey also stresses the role played by farmers in promoting erosion and loss of soil fertility within the Lake Titicaca region. It claims that over-grazing has completely obliterated the vegetation cover in some localities; sheep have compacted soils, destroying their structure and capacity for filtration. Also damaging have been the practices of burning field stubble and 'heavy' ploughing.

Although only a handful of small farmers in the study region have the financial resources to purchase second-hand tractors (one Chuan *campesino* was in 1996 in the process of buying a tractor over a three years period at a cost of US $ 25,000, pledging his house and land as security), tractors can be hired by the hour from a naval base, Huatajata and Huarina. Whilst the 1996 'strategy for the transformation of agricultural productivity' (p.96) advocates 'a technological leap' in farming, it is abundantly clear that nothing is contributing more to soil deterioration in some communities than the tractor. Apart from exposing sizeable areas of fragile, over-worked soils to the powerful erosive effects of storms and strong winds, the use of tractors in small fields is totally inappropriate: in Chua Visalaya on one occasion in 1996 a tractor was observed reversing after ploughing each furrow because the minute size of the plot made turning round impossible. Additionally, *campesinas* complain bitterly about the back-breaking work (often leading to rheumatism) involved in breaking up large, compacted clods of earth with their heavy *chontas* - one of the reasons given by teenage girls for not wanting to work on the land. Over recent years a few of the younger women have indicated that they might be more prepared to remain in the region (combining their city marketing transactions with farming) if more appropriate technology labour-saving devices, e.g. ones to ease the tasks of winnowing and grinding grain, were to be introduced.

One of the limited number of successful technological innovations in farming within the region during recent decades evolved in the mid-1980s, after being introduced by NGOs in the low-income housing areas of El Alto. *Carpas solares* (solar greenhouses), built cheaply of polythene on *adobe* foundations, take advantage of strong solar insolation during the day, whilst providing protection from damaging frosts at night. Although lettuces, radishes, tomatoes, carrots etc. added variety to local diets and provided a source of extra income, most *carpas solares* in lakeside communities were abandoned once NGO support was withdrawn. They are now making a 'come-back': a number of hoteliers are constructing sophisticated examples, having seen the advantages of being able to provide clients with freshly grown, uncontaminated salad vegetables.

Discontinuing traditional fallow practices over the years and, in some cases, adapting rotation systems to enable extra crops of onions and/or beans to be grown, have undoubtedly contributed to the general reduction in soil fertility. Though most houses within the region nowadays contain kerosene stoves or electric cookers (as in 1981, often provided by migrant children working in urban areas) and no longer require animal dung for heating or cooking purposes, the manure thus made available for usage as organic fertilizer fails to meet the needs of ever-more intensive cultivation, especially as the numbers of animals kept by many *campesino* families have been reduced.

According to a 1998 Ministry of Agriculture report, 'Bolivia is by far the lowest consuming country of fertilizer nutrients per cultivated hectare in Latin America' (*BT*). Since all chemical fertilizers still have to be imported, *campesinos* are obliged to pay 40 to 60 per cent more than is the case in neighbouring fertilizer-producing republics. It is to be hoped that the situation will greatly improve once plans for building a factory in the eastern lowlands, to produce *Urea* (an inorganic fertilizer made from natural gas, water and nitrogen) materialize. Although there is no shortage of *Urea* from Chile, Peru and Argentina on sale in Huatajata's Wednesday market, few *campesinos* can afford to pay approximately

US $ 25 per sack; they are more likely to purchase several kilos and apply the fertilizer in inadequate quantities to fields planted with potatoes and *oca*.

Despite the low night temperatures at such a high altitude, lakeside crops, especially potatoes and *oca*, are subject to infestation by grubs and insects; in both 1996 and 1998 harvests were more than halved in some communities by grubs. The spread of crop diseases is facilitated by the tendency for poorer *campesinos* to plant already diseased roots or seeds, or to cultivate crops on contaminated land. Lacking information and advice, many *campesinos* continue to confuse fertilizers with insecticides and pesticides. Sadly, several of the insecticides and pesticides readily available for purchase at local weekly markets are dangerous to apply, banned by many governments and only on sale as a result of contraband activities. Principal amongst these is *Aldrin*, banned by the Bolivian government; not only does it contaminate the soil but it also leaves a toxic residue in the crop foliage.

ALT's strategic plan maintains that problems of contamination and pollution around Lake Titicaca are confined to the Puno region. Whilst Puno's water contamination problems are certainly acute - as is only too apparent from the presence of the thick blanket of green algae completely surrounding the port - pollution and contamination problems are rapidly spreading throughout the region.

> In effect, many rivers especially Peruvian rivers, arrive already contaminated with organic, detergent, and industrial wastes and other damaging products, resulting in the lake's much-feared artificial contamination ... The Sacred Lake of the Incas has become a convenient dumping ground where anybody can drop off highly contaminated liquids (Vargas, 1996).

Increasing levels of water contamination, together with indiscriminate fishing, spell imminent disaster for already dwindling lacustrine fish populations. During the protests in 1996 against the proposed *Ley INRA*, the sub-secretary of Forestry and Fishing Development publicized the fact that *pejerrey* (originally introduced to Bolivia from Patagonia in 1957), accounting for 45 per cent of the fish consumed by Paceños, was one of a number of Lake Titicaca fish species on the verge of extinction as a result of 'mismanagement'. Countless of 'the lake's 16,888 fishermen, largely dependent on fishing for subsistence', or supplementing their meagre earnings from farming, would be 'gravely affected'.

If contamination remains uncontrolled it is also likely to have a seriously damaging effect on *campesino* livestock, partly reliant on forage in the form of lake-grown *totora* and *chancco*, and on lakeside fields used for cultivation purposes. The ever-growing problems of pollution in the study area are dramatically illustrated by Peter McFarren in a 1996 *BT* article entitled 'The sacred spots of Lake Titicaca and Copacabana turn to rubbish':

Pollution is seen all along the road from La Paz to Copacabana. ... the shores of Tiquina are full of beer cans, plastic bags and litter ... All along the shores, the drains of the communities, restaurants and hotels lead straight into the lake ... each week people from La Paz who visit the shores of Lake Titicaca by bus leave hundreds of kilograms of plastic behind them.

Examples of cans and bottles thrown from passing vehicles later causing injury to animals are growing in number. In some waste areas, such as on land surrounding derelict houses, *campesinos* have themselves added to the pollution and health hazards by emptying rubbish on the ground and even leaving animal carcasses to rot.

It has to be said that, whilst the Aymara *campesinos* living in the lake region have not relinquished their reverence and respect for *Pachamama*, unfortunately for a number of years this has not been reflected in their attitudes towards soil erosion and pollution. Today there appears to be little recognition that human activities are partially responsible for land degradation - or that there is any need to make every effort to preserve soil fertility for future generations. Rarely is any concern expressed about the detrimental effects of over-grazing or the implications of leaving deeply ploughed fragile soils exposed to the elements and vulnerable to severe gully erosion. Poor harvests are beyond farmers' control i.e. they are due to inclement weather (often a contributory factor) or *plagas* (plagues of grubs or insects), which many still fervently believe represent the revenge of *Pachamama* for some offence committed by a member of the community.

Ley INRA has failed to offer any solution to the complicated question of water rights - an issue as controversial as land ownership on the arid Altiplano. In the Lake Titicaca region the increased demand of *campesinos* for water to irrigate cash crops of onions and beans has intensified inter-family and inter-community antagonism. As previously mentioned (p.106), competition for limited water resources has taken on a new dimension in the 1990s. A growing number of *campesinos* able to pay approximately US $ 500 for a motorized water pump have few qualms about depriving their less fortunate neighbours of fresh water for irrigation purposes; some poorer *campesinos* comment that this type of abuse creates feelings of helplessness and strengthens their resolve to 'leave the land'.

A growing cause for concern at the turn of the century is the increasing rate of salinization in soils near the lake, mainly resulting from the extraction of lake water for irrigation purposes. Whilst lake floods obviously leave salt deposits on their retreat, water from the lake was rarely applied deliberately to fields in former times; even in the early 1980s the majority of lakesiders considered the water too sacred to extract - to be drowned in the lake was regarded as 'an honour' and 'a privilege'. For the moment *campesinos* remain divided about the relationship between salinity and productivity: some insist that saline soil creates no problems whilst others are adamant that 'it burns plants'. Environmentalists' fears about the increased pressures on lake water have even elicited articles on 'the need to avoid an Aral'! Proposals made by NGOs in the early 1980s to construct a system of small dams and irrigation channels in the mountains above the lakeside communities are still discussed by community leaders but it is unlikely that they will materialize because of the cost and the labour essential for the success of such a venture.

Stagnant water remains a formidable health hazard to humans and animals alike. According to vets working at the Belén experimental station, *distomatosis* (locally known as *jhalphalaka* and caused by *fasciola hepática*) constitutes one of the three main health problems affecting animals within the region, the others being respiratory problems and malnutrition. Over recent years the disease has exacted a heavy toll in terms of cattle, pig and sheep losses, partly because of the withdrawal of departmental fumigation services. Although some *campesinos* are aware of the causes of so many animal deaths (and give that as one of the reasons for burning fluke-infested *totora*), few can afford to pay US $ 8 per annum per animal for medical protection.

None of the major agricultural problems identified by lakeside *campesinos* in 1996 (p.84) were resolved by *Ley INRA*. The credit facilities and extension services made available to *campesino* communities by the 1953 Agrarian Reform Law, and withdrawn in the 1980s, have not been restored: there is no comparison between the support given to communities in the 1970s and that received in the late 1990s. Whilst incomes derived from farming activities have fallen - by as much as 50 per cent according to some farmers - there is little prospect of the situation improving dramatically in the near future. The ALT 1995 strategic plan alleges that 'the *campesino* in general does not seek to maximize production, rather to diminish the risks'. From his point of view there is every reason why he should act in this way since the odds are heavily weighted against him.

Apart from labour shortages and the dread of crop diseases, the *campesino* has no control over the weather and in recent years has suffered considerable financial loss as a result of the vagaries of *El Niño*. In 1998 *campesinos* were even beginning to discuss the implications or 'global warming'. For the first time in living memory June frosts failed to occur, making it impossible for the potatoes and *oca* not destroyed by grubs to be dehydrated by the freeze-thaw mechanism. Farmers receive no compensation for poor harvests due to 'natural disasters'; instead they are forced to sell stock at minimal prices, to borrow money from city-based relatives or to boost their income by finding temporary low-paid manual work wherever possible.

Since the closure of the Agricultural Bank in 1985, obtaining credit for farming needs (seeds, tubers, animals, medicines, tools, fertilizers etc.) has proved extremely difficult for *campesinos*. A few banks which claim to assist small farmers are normally reluctant to risk making even small loans to individuals with insecure incomes: in any case, their high interest rates dissuade *campesinos* from pursuing the issue. Several banks have during the last decade made credit available on an escalating scale to small groups of rural-urban migrants (including onion vendors) but impose stringent conditions e.g. immediate termination of the contract if one member of the group fails to make a monthly payment on time. Under the terms of the 1994 Law of Popular Participation (described by Sánchez de Lozada as signifying 'the most important redistribution of political and economic power in the republic since the 1952 revolution'), provincial officials assumed the responsibility for distributing public funds to individual communities. Whilst lakeside communities had hoped that such funds could be used partly for agricultural purposes, a number, including Huatajata, have been forced to resort to threats of legal action against municipal authorities refusing to release allocated allowances.

In addition to all these difficulties, *campesino* producers face the formidable marketing problems discussed at the end of the last chapter. Lakeside farmers have failed almost entirely to respond to dramatically changing marketing and financial conditions. Unfortunately they have been given no expert advice about adapting to new situations and on appropriate innovative changes e.g. in crop specialization. Onions, which in the 1970s and early 1980s provided a steady income for communities such as Llamacachi, are nowadays subject to strong competition from those produced under more favourable climatic conditions in the Cochabamba and Yungas regions. Lakeside producers allege that they have to produce **five times** as many onions today to make the same profits as in former times. Simultaneously food donations are undermining the production of traditional crops. Whereas in 1970, 6,000 tons of donated food entered the country, by 1990 the volume had risen to 290,000 tons, much of it US aid, in the form of wheat and derivatives of wheat and milk. Such commodities are sold in urban markets (to provide funding for government social projects), competing directly with *campesino* produce. According to a report prepared in 1995 for the Inter-American Bank, in some parts of the country donated food accounts for up to 40 per cent of the *campesino* diet.

At the same time, partly as a result of the liberalization of trade and lack of protection given to locally-produced crops, the diets of urban populations are undergoing more radical change than at any time since the Spanish Conquest. According to the national executive of *Alimentaria Seguridad* (Pensioner Security, 1996) 'bread, sugar, pasta and rice' have become 'the most consumed products in urban centres', largely displacing traditionally consumed roots and cereals. Even in the Lake Titicaca region, evening meals of *chairo* (*chuño* soup) or *quinua* soup are being replaced by soups made from maize, rice and pasta. In such circumstances - at a time when one of the main ambitions of more than a few teenagers in La Paz and El Alto is to be able to afford to patronize the new McDonalds on a regular basis - there is no incentive whatsoever for elderly *campesinos* to use their limited cash resources to buy improved seed potatoes or chemical fertilizers, in order to produce food surplus to family requirements for sale in urban markets. What seems vital to the survival of *campesino* communities is a shift in emphasis in terms of crop and/or animal production. The time has come to respond positively to new trends and changing food demands. Had the government's extension services remained in operation, and progressively minded young *campesinos* not abandoned the countryside in search of more lucrative opportunities, such a change might well have occurred some years ago; as it is, agricultural production within the region has been allowed to 'stagnate'.

The impact of rural-urban migration on farming communities is incalculable. Many elderly *campesinos* regard today's labour shortage as their greatest problem and believe it poses an insuperable threat to future farming. A number of old couples are convinced that, whatever their offspring say about holding on to family land, they will sell it at the earliest opportunity i.e. once their parents are dead. Whilst migration was a well established trend and already presenting difficulties in some families in the early 1970s, it has now reached an alarming stage, as what appears to be the last generation of farmers literally 'dies out'. In 1998, of 10 houses in a row in Llamacachi, only one was permanently occupied, several had been deserted and the others were only being used on occasions by migrants attending *fiestas*.

The 'push' and 'pull' factors sustaining migration over the years to La Paz and El Alto (whose population of 405,000 in 1992 had risen to 700,000 by 1997) are multifarious and complicated. 'There's nothing left here', 'there's no money', 'life is too hard in the country', 'there's no future in farming', 'there's plenty of work in the city' and 'there's more life there' epitomize the feelings of young people eager to start a new life in an urban environment. In *Forever in your debt? Eight poor nations and the G-8* (1998), the NGO, Christian Aid, paints a very dismal picture of the reality of life for rural migrants to El Alto:

El Alto in Bolivia, a former squatter settlement of rural migrants, is today one of the fastest growing cities in Latin America. In the 1980s, as a result of the IMF's drive to cut public expenditure and nationalise many state-owned companies, a steady stream of migrants turned into a flood; thousands of miners were made redundant and subsistence farmers, faced with increasing competition from cheap foreign import crops, were no longer able to eke out a living from their fields. They came to El Alto, full of expectations, hoping for a chance to secure work and a better education for their children. However, their dreams have been left unfulfilled ... El Alto is generally seen as a 'no-go-area', characterised by rocketing crime rates which mirror the rise in long-term unemployment and frustration of the young people, who make up 65 per cent of the city's population.

The majority of migrants from the lakeside fare better. They share the advantage of familiarity with the cities of El Alto and La Paz and are likely to start their urban lives lodging with relatives; some of the young women establish lasting friendships and entrepreneurial links as weekend vendors of onions. A number of lakeside dwellers have prospered since migrating. For example, a Llamacacheño, known to the writer as a school boy in 1971, presently runs an ethnic clothes business in El Alto, employing 25 workers and supplying tourist shops in La Paz, France, Germany and Israel. By 1991 a Chuan migrant had become the president of her *zona*, the first woman to be president of any sector of El Alto. Such examples have encouraged other migrants with entrepreneurial and political ambitions to seize all available opportunities to improve their situation.

The departure of energetic, potentially innovative, educated young men and women from the Lake Titicaca region has had a dramatic effect on traditional farming practices and gender roles. Because of acute labour shortages, elderly women, who would previously have spent much of their time herding animals on the hill slopes, are obliged instead to carry out cultivation tasks. As a direct result many families' flocks of sheep have been considerably reduced in size; those remaining are generally grazed near fields being worked, with the inevitable consequences in terms of soil erosion. Additionally women are occasionally seen ploughing with teams of oxen - an unthinkable occurrence in the early 1970s.

During the last decade a few rural-urban migrants - deserted wives and elderly couples wanting to live out their days in the 'more peaceful' communities of their birth - have returned to the lake region. However it is very difficult to envisage any circumstances capable of enticing the mass of rural-urban migrants to return to the area, except possibly the establishment of small-scale industrial enterprises. Certainly rural education has done nothing to encourage school leavers to remain in the countryside; on the contrary, it has raised their expectations. The higher quality of urban education has in the past persuaded some *campesinos* to have their children educated in city schools (whilst staying with relatives), easing their way into urban living. Regrettably, a number of migrants have become ashamed of their rural origins. One young woman originally from Llamacachi was heard to remark in 1998: 'it's a great pity that my mother is a *campesina*'.

It is often said by outsiders that at least rural-urban migration has eased the problem of excessive population pressure on land resources - but at what cost? At the same time communities have continued to lose land in a variety of ways. A significant amount of land *in toto* has been lost by replacing old *adobe* houses with larger brick buildings, leaving the former dwellings to the forces of nature. Although most families have welcomed road improvements (including road widening), which have been largely responsible for halving the previous travel time to La Paz, they have been less enthusiastic about the excavating of several dry river beds to obtain gravel for road maintenance; some of the sites have become eyesores, detracting from the general beauty of the area. One community not benefiting from road improvements is Huarina: largely because of the construction of a by-pass, the old settlement has taken on the appearance of a ghost town. Its market and shops have closed whilst the bulk of the population have abandoned what was formerly a thriving urban centre.

Most lakeside dwellers have adopted an ambivalent attitude towards tourism: tourists generally pass through farming communities without stopping. To Huatajata's farmers tourism has proved a mixed blessing. Whilst restaurants and hotels have been built on what was previously cultivable land, opportunities for diversification have averted large-scale migration. *Campesinos* are able to supplement farm incomes by supplying restaurants and local markets with freshly caught fish (the community's fishing cooperative now has more than 60 members) and by running boat excursions to a number of lake islands. Huatajata's future seems more secure than that of any other community between Huarina and Tiquina. Significantly, at a distance of 87 km from La Paz, Huatajata is becoming recognized as a commuting centre - a new phenomenon in this part of Bolivia. According to the community's elders more than 50 Huatajatans are now operating as commuters i.e. residing in the community whilst travelling on a daily basis to work in La Paz or El Alto. On the other hand, whilst at least five Huatajatan teachers work in urban schools, a much larger group from La Paz and El Alto make the daily journey to schools in Huatajata and even Compi.

Has the passage of time erased the former differences between the communities of Llamacachi and Chua Visalaya? According to Llamacacheños, proud of their new hotel (built on a plot of land sold by a migrant on his father's death), their superiority remains indestructible. They readily acknowledge that they have been affected to a greater extent by rural-urban migration, but are eager to confirm that most migrants from the community have found lucrative openings. Llamacacheños have no doubts about their business acumen and continuing superiority as entrepreneurs: unlike their *ex-colono* neighbours, they would **never** allow **their** daughters to enter domestic service in the city - they would be running their **own** businesses. On the other hand, it is clearly apparent that they are resentful and envious of the fact that Chua has recently gained the title of cantonal centre: Chua Visalaya has acquired a *plaza*, around which several stores have been built. Its status has also been enhanced by the construction of a naval school's landing pier on the Chuan water front. *Campesinos* in Chua Visalaya allege that, since the disbanding of the community's cooperative in 1973, there has been more harmony and cooperation than in Llamacachi, where age-old land disputes have not been resolved by *Ley INRA* and religious differences continue to divide the community. They argue that, unlike Llamacachi, Chua Visalaya has its own community primary and secondary schools and that any shortcomings there might have been in the past, due to their lack of marketing experience as *colonos*, have long since been remedied.

Proximity to La Paz and greatly improved communications have played a decisive role in the study region's 'development' since the National Revolution. In more isolated areas, life in many farming communities continues much as it has done since the early days of agrarian reform. Whereas it is not unknown nowadays for a queue of 50 vehicles to be held up by a *fiesta* procession in communities such as Huatajata and Compi, in the municipal centre of Ancoraimes, on an alternative route to southern Peru along the eastern margins of the main lake, it is not unusual for visitors to have to wait from mid-day to sunset for transport to La Paz. Easy access to the city from the north-eastern shores of Huiñaymarca has undoubtedly facilitated the expansion of *campesinos'* marketing activities, in addition to their acquisition of consumer goods: it has also had an important bearing on lakeside dwellers' decisions to migrate. It is more than conceivable that it may ultimately be held responsible for the complete collapse of community farming.

With growing demands for recreational space and concern for a better 'quality of life', a number of wealthy Paceños are beginning to re-appraise their life styles. At the end of the century, the city's air pollution problems, attributable mainly to ever-increasing quantities of vehicle emissions, are assuming significant proportions. Inhabitants of La Paz are also exposed to polluted air drifting down into the city's basin from the domestic waste, uncontrolled industrial enterprises and congested traffic of El Alto, which has the unenviable reputation of being the republic's most polluted urban centre. Today's easy access to the scenically attractive lakeside makes it an ideal location for second, weekend and holiday homes; the possibility of commuting to La Paz on a daily basis presents *Paceños* with the option of uprooting themselves entirely and settling permanently beside the lake - providing they can secure suitable plots of land on which to build housing. Even before *Ley INRA* was passed a number of land plots had been sold by *campesinos* to ex-politicians (including an ex-president), military personnel, doctors and journalists. Some of the recent purchases remain surrounded by large, obtrusive walls and fences, whilst others have already been built on.

When fixing land sale prices in the Lake Titicaca region, the Agrarian Superintendency currently distinguishes between land adjacent to the lake and land above the through-road. In 1998 the most desirable land between the lake-shore and the road was valued at US \$25 per sq m, whilst that above the road was priced at US \$15 per sq m. What represents a modest sum to speculators and well endowed Paceños is considered a fortune by elderly *campesinos*, struggling to cultivate their fields, and a very welcome legacy by rural-urban migrants with no intentions whatsoever of returning to live in the area. Sadly, it is beginning to seem likely that within the next 20 years most of the land in question will be sold to outsiders and farming virtually disappear from a part of Bolivia where it has survived the turmoils of invasion, massacre, colonialism, rebellion and land seizure over many centuries.

Abbreviations used in the text

ALT	Lake Titicaca Binational Authority
ANAPQUI	National Association of Quinua Producers
BT	*Bolivian Times* (published in La Paz)
CAN	National Agrarian Reform Commission
CEDOIN	Centre of Documentation and Information, La Paz
CIDOB	Indigenous Confederation of Eastern Bolivia
CNRA	National Agrarian Reform Council
CNTCB	Bolivian National Confederation of Campesinos
COB	Bolivian Workers' Central (Union)
COMIBOL	Mineral Corporation of Bolivia
CORDEPAZ	La Paz Department Development Corporation
CSCB	Confederation of Bolivian Colonizers
CSUTCB	Confederation of Bolivian Campesino Workers
ECLA	Economic Commission for Latin America (United Nations)
EMI	Military School of Engineering, La Paz
FAO	Food and Agriculture Organization (United Nations)
HISBOL	Bolivian Institute of Social History
IMF	International Monetary Fund
INC	National Colonization Institute
INKA	National Institute of Kollasuyo and Amazonia
INRA	National Agrarian Reform Institute
INTI	National Land Institute
LIDEMA	Alliance for the Defence of the Environment (NGO)
LP	*Latinamerica Press* (published in Lima)
MACA	Ministry of Peasant Affairs and Agriculture
MNR	National Revolutionary Movement (Political party)

NCDP	National Community Development Programme
NGO	Non-Governmental Organization
ORSTOM	French Institute of Scientific Research for Development and Cooperation
PELT	Special Lake Titicaca Project
PIR	Party of the Revolutionary Left
POR	Revolutionary Workers' Party
PROSUKO	Inter-Institutional Program of Suka Kollus (NGO)
SEMTA	Multiple Services in Appropriate Technology (NGO)
SNRA	National Agrarian Reform Service
TDC	'Village Level Worker', employed by NCDP in association with US Peace Corps
TDPS system	Enclosed basin of the Altiplano containing Lake Titicaca, the River Desaguadero, Lake Poopó and the Coípasa Salt Lake
UMSA	University of Saint Andrews, La Paz
VSF	Veterinarians Without Borders (NGO)
YPFB	National Oil Corporation

Glossary

Sp - Spanish; Aym - Aymara; Qu - Quechua

adjudicación (Sp): the process of adjudication usually leading to the sale of land for a price fixed by the Agrarian Superintendency (*Ley INRA*)

adobe (Sp): literally an unburnt, sun-dried brick, but generally denoting a house built of *adobe* bricks

agregados (Sp): land-owning, but not 'original', residents of a *comunidad originaria*

agua potable (Sp): drinkable water

aini, ayni (Aym): system of reciprocal labour, lending and borrowing

aljería (Sp): *hacienda* warehouse and shop in the city

altiplano (Sp): Andean high plateau or plain

andenes (Sp): pre-Columbian walled terraces

anticrético (Sp): arrangement whereby creditors enjoy usufruct rights to certain of their debtors' fields in lieu of receiving financial payment

arroba (Sp): unit of weight (approximately 11.5 kgs)

asentamiento (Sp): Chilean cooperative farm

avena (Sp): oats, introduced to the Andean region by Spanish *hacendados*

awatiri (Aym): *hacienda* shepherd

ayllu (Aym/Qu): clan or kinship system, with close social ties and collective ownership of land

aynoka, aynoqa (Aym): community lands cultivated on an individual basis but in strict accordance with a prescribed rotation system

balsa (Sp), *yampu* (Aym): boat constructed of *totora* reeds

barrio, barriada, zona (Sp): low-income housing quarter

bombas de agua (Sp): motorized water pumps used for irrigation

brujo (Sp), *yatiri* (Aym): sorcerer, magician

burro (Sp): donkey, highly valued as a beast of burden in the lakeside region *camana* (Sp): guardian or supervisor e.g. *camana de productos*, the person assigned to make *chuño* on the *hacienda*

campesino/a (Sp): countryman/woman, peasant farmer, officially replacing the pejorative term, *indio*, after 1952

cañahua/hui, qanawi (Qu): small-grained, nutritious cereal, native to the Altiplano and able to grow up to 4,000 m

cantón (Sp): smallest administrative unit in Bolivia's local government system

carpas solares (Sp): solar greenhouses, built of *adobe* and polythene, introduced to the Lake Titicaca region in the 1980s

casa hacienda (Sp): estate house, residence of *hacendado* and family

casa de ovejas (Sp): *hacienda* 'sheep house'/barn

caya, c'aya (Aym): *kaya* (Qu): frozen and dried *oca*

cebada (Sp): barley, introduced as a crop by Spanish landlords; *cebadilla* (Sp): wild barley

chacra (Sp): in the Andean region *chacra*, like *kallpa* (Aym), refers to a small field

chairo (Aym): *chuño* soup

chálla, ch'alla (Aym): the spilling of alcohol on the earth, as a token of respect for *Pachamama*, before a celebration or task

chalona (Aym): dried, salted meat

chancco, chanko (Aym): aquatic plant traditionally used for forage

charqui (Sp), *ch'arki* (Aym): meat cut into strips and dried in the sun (the term from which 'jerked meat' is derived)

chicha (Sp): beer made from masticated maize flour, fermented and boiled

chonta (Aym): clod breaker used mainly by *campesinas*

chuño, ch'uñu (Aym): dehydrated potato (black in appearance)

cocaleros (Sp): growers of coca for both traditional uses and cocaine

colono (Sp): peasant or serf working for an estate owner on the land and fulfilling other obligations in pre-reform times, in return for usufruct rights to small land plots

comunario (Sp): peasant living in a *comunidad indígena* (Indian community) or

comunidad originaria (Sp): (a freeholding), not subject to the *hacienda* system in the pre-reform period

conquistadores (Sp): Spanish conquerors of the late fifteenth and early sixteenth centuries

cordillera (Sp): mountain chain or ridge; *serranía* is a term similarly used

corregidor (Sp): in *hacienda* times, the *mestizo* who administered a *cantón*; later, aide to *intendente*, the highest authority in the *cantón*

curandero (Sp): curer, using herbal remedies

demanda (Sp): a land claim petition presented to the SNRA in accordance with 1953 legislation

(en) descansco (Sp): 'resting' i.e. fallow period

distomatosis (Sp), *jhalphalaka* (Aym): riverfluke disease causing high mortality levels in livestock

dotación (Sp): process of land settlement, granting land to *campesinos* and indigenous peoples without payment: *dotado*, recipient of land grant

ejido (Sp): collective farm unit in Mexico

empresa agricola (Sp): agricultural enterprise (1953 Agrarian Reform Decree): replaced by *empresa agropecuaria* in *Ley INRA*

empresa ganadera (Sp): capitalized cattle ranch/agricultural livestock enterprise

empresario (Sp): commercial farmer

encomienda (Sp): royal land grant in Spanish colonial period

estancia (Sp): sub-division or section of an *hacienda*

expediente (Sp): dossier compiled by the SNRA as a preliminary to land redistribution, containing accumulated documents relating to property (1953 legislation)

expropiación (Sp): according to *Ley INRA*, land expropriation involving the payment of compensation

feria (Sp): weekday market: *feria franca*, open market in urban area

fiesta (Sp): festival, holiday

forastero (Sp): 'foreigner', landless *comunario*

guacchallama (Sp): collective term for flocks of llamas and alpacas

habas (Sp): broad beans, an important crop in lakeside communities

hacendado (Sp): landlord or *patrón*

hacienda (Sp): landed estate cultivated by a system of serfdom (*colonato* system) before 1953

huracán (Sp): freak wind, hurricane

ichu (Qu), *paja brava* (Sp): stiff, spiky grass native to the Altiplano, used for pasturing camelids

isañu (Aym): an edible tuber traditionally grown between 3,000 and 4,000 m above sea level

jilakata (Aym): traditionally the highest official in Indian communities; in *hacienda* times the *jilakata* was the task master directing *colonos'* work on estate land

kuraka, khuraka (Aym): Aymara regional chief in pre-colonial days

latifundio (Sp): extensive, unproductive estate; *latifundista, latifundio* owner

loteadores (Sp): term applied to individuals seizing tracts of land illegally, especially in the Santa Cruz region

llanos (Sp): plains, in Bolivia usually indicating extensive grasslands

'maria' (Sp): name given in lakeside communities to a *balsa* with only one pointed end

mayordomo (Sp): *hacienda* administrator or overseer, usually a *mestizo* or a *'blanco'*

mayoruni (Sp): *colono* working one day weekly for usufruct rights *mestizo* (Sp): person of 'mixed' blood, usually Indian and Spanish

minifundio (Sp): a very small landholding, farmed by a *minifundista:* *minifundismo* refers to the excessive fragmentation of landholdings

'la miseria' (Sp): term applied in Altiplano communities to the period of hardship and food shortage from November to April

mita (Sp): colonial obligation of Indians to work in the mines, later extended to agricultural tasks

mitani, mit'ani (Sp): duties relating to cooking and cleaning in the *hacienda* house; sometimes applied to the person undertaking the duties

neolatifundismo (Sp): the re-emergence and consolidation of the *latifundio*, supposedly abolished by 1953 legislation

oca, oqa (Qu): root vegetable native to the Altiplano; in pre-reform times *oca* was second in importance to *papa* (the potato)

originario (Sp): *comunario* i.e. descendant of an original settler in a *comunidad originaria*

orilla (Sp): lake-shore

Pachamama (Aym): Mother Earth, venerated by Aymara and Quechua people alike

papalisa, papa lisa (Sp), *ullucu* (Aym), *ulluku* (Qu): small, sweet edible tuber

páramos (Sp): high, cold, treeless basins and valleys in the Andean region

parcela (Sp): parcel or plot of land; *parcelero*, the owner of a number of small land plots

patrón (Sp): *colonos'* term for their landlord

plagas (Sp): usually referring to crop 'plagues' (diseases), especially grub and insect infestations

persona, media persona (Sp): categories of *colonos* with access to land plots according to their labour input

pongueaje (Aym derivation): personal and domestic services which *colonos* were obliged to render to the landlord and his family

preste (Sp): the most important sponsor in the *fiesta* system

propiedad agraria cooperativa (Sp): cooperative property; *propiedad pequeña*, smallholding

pueblos indigenas (Sp): indigenous peoples, usually in Bolivia referring specifically to forest-dwelling ethnic groups

puna (Sp): cold, arid, inhospitable area between 3,000 m and the permanent snow-line, vegetated by *ichu* grass

quinua/oa, kiuña (Qu): millet-like cereal of high protein content, native to the Altiplano (the staple diet of the Inca people)

reforma agraria (Sp): land and agrarian reform; *contrarreforma*, counter-reform

resolución suprema (Sp): supreme resolution granted by the CNRA and authorizing the implementation of land redistribution within an *ex-hacienda*

reversión (Sp): land reversion, one of three components of *Ley INRA's* land distribution rationalization programme

saneamiento (Sp): process of eradicating landholding irregularities (1996 legislation): a property 'cleansed' by *saneamiento* is considered *sano*

sayaña (Aym): traditionally a houseplot; term sometimes applied by *campesinos* to all but *aynoka* plots

selva (Sp): tropical rain forests of northern and eastern Bolivia

sindicato (Sp): peasant union, community-based syndicate

solar campesino (Sp): *campesino* ground plot/residential plot referred to in 1953 agrarian reform legislation

suka kollus (Aym), *waru waru* (Qu), *camellones* (Sp): ancient raised field systems

surcofundismo (Sp): extreme fragmentation leading to the cultivation by individuals of single furrows in a number of land plots

taclla, taqlla (Qu): wooden foot plough used since Inca times

tambo (Qu/Sp): Inca state warehouse (for food storage): today, an open market in an urban centre, serving as a distribution point where truckers can unload goods directly

tarwi, tarhui (Aym): a lupin plant traditionally grown in lakeside communities and used as a herbal remedy for diabetes, rheumatism etc.

tormenta de graniza (Sp): hail storm

totora (Aym): lake reeds formerly used for making *balsas*, as animal fodder, for thatching, bedding and fuel; *totorales, totora* beds along lake margins

trigo (Sp): wheat, a cereal introduced by the Spaniards

tunta (Aym): dehydrated potato (white in appearance)

yanaconas (Aym derivation): landless people, possibly slaves in pre-colonial times

Yungas (Sp): semi-tropical, deeply incised valleys of the Cordillera Real

Bibliography

Abbott, J.C. and Makeham, J.P. (1990, 2nd ed.), *Agricultural Economics and Marketing in the Tropics*, Longmans Group: UK.

Alanes, Z. (1993), 'Un debate que regresa: La Reforma Agraria cuarenta años después', *La Razon* (La Paz), 23 May.

Albó, X., Godínez, A., Libermann, K. and Pifarré, F. (1989), *Para Comprender las Culturas Rurales en Bolivia*, MEC, CIPCA and UNICEF: La Paz.

Alexander, R.J. (1958), *The Bolivian National Revolution*, Rutgers University Press: New Jersey.

Alexandratos, N. (ed.) (1995), *World Agriculture: Towards 2010*, FAO and John Wiley and Sons: Chichester, UK.

Andean Group Report (1998), *Latin American Regional Reports*, Latin American Newsletters: London, 6 October.

Antezana E., L. (1971), *El feudalismo de Melgarejo y la Reforma Agraria en Bolivia*: La Paz.

Antezana E., L. (1979), *Proceso y Sentencia a la Reforma Agraria en Bolivia*, Puerta del Sol: La Paz.

Antezana E., L. and Antezana S., A. (1997) *Juicio y Condena a la Ley INRA*, Fondo Editorial de Diputador: La Paz.

Aruquipa Z., J.A. (1997) Articles in *Bolivian Times* (La Paz) 'All Time is Relevant and Flexible' (16 January); 'The Gods of the Aymara Religion: Aymara Introspective' (23 January).

Autoridad Binacional Autonoma del Lago Titicaca (1997), *Cuenca: Titicaca-Desaguadero-Poopó-Salar*, CIMA: La Paz.

Banco Agrícola de Bolivia (1964) , *La Cooperativa Agrícola Chua Limitada*: La Paz.

Banco Agrícola de Bolivia (1970), *La Ganadería Boliviana: Situación de la Comercialización de la Carne y Posibilidades para su Desarrollo*: La Paz.

Barke, M. and O'Hare, G. (1991), *The Third World: Conceptual Frameworks in Geography*, Oliver and Boyd, Longman House: London.

Baumann, G.F. (1970), 'The National Community Development Programme in Bolivia and the Utilization of Peace Corps Volunteers', *Community Development Journal*, Vol. 5, No. 4., October.

Bebbington, A. (1998), 'Sustaining the Andes? Social Capital and Policies for Rural Regeneration in Bolivia', *Mountain Research and Development*, Vol.18, No.2.

Bebbington, A. and Thiele, G. (1993), *Non-Governmental Organizations and the State in Latin America*, Routledge: London.

Benton, J.M. (1974) *Some aspects of change in post-revolutionary Bolivia: a geographical study of Aymara communities beside Lake Titicaca*, Ph.D. Thesis: Keele University, UK.

Benton, J.M. (1983), 'Development on the Lake Titicaca shore', *Geographical Magazine*, Vol.LV, No.2. February.

Conference papers presented: *Domination and stagnation: the impact of the hacienda on the Bolivian rural scene* (at an international symposium on 'Landlord and Peasant in Latin America', Cambridge University, 1972); *Conflicting values in community change: some reflections on rural development in Bolivia's Lake Titicaca region* (at an Anglo-Mexican conference in Mexico City, 1983); *The impact of national politics and the monetary system on agricultural development* (at an Anglo-Mexican seminar, Oxford University, 1986); *Food procuring strategies adopted by low income urban women in Bolivia* (at an International Geographical Union Conference, Rutgers University, 1992).

Blakemore, H. (1966), *Latin America*, Oxford University Press: Oxford.

Blakemore, H. and Smith, C.T. (eds.) (1971), *Latin America: Geographical Perspectives*, Methuen: London.

Bolivia Bulletin, formerly a bimonthly publication of CEDOIN: La Paz.

'Tapping the Source: Water management in Bolivia' (October (1989); 'How Green is the Valley: a second look at Bolivia's ecology' (April 1991); '500 Years: The struggle against internal colonialism' (April 1992); 'The Fate of Agrarian Reform' (December 1992); 'Focus on Indigenous - Defence of their Territory: The struggle goes on' (February 1996).

Bollinger, A. (1993), *Así se alimentaban los Inkas*, Los Amigos del Libro: La Paz.

Bouysse-Cassagne, T. (1991), 'Poblaciones humanas antiguas y actuales', in Dejoux, C. and Iltis, A. (eds.), *El Lago Titicaca: Síntesis del conocimiento limnológico actual*.

Box, B. (ed.) (1998), *South American Handbook*, Footprint Handbooks: Bath, UK.

Brandt, W. (chairman) (1980), *North South: A programme for survival* (Report of the Independent Commission on International Development Issues), Pan Books: London.

Buechler, H.C. and Buechler, J-M. (1971), *The Bolivian Aymara*, Case Studies in Cultural Anthropology, Holt, Rinehart and Winston: New York.

Buechler, J-M. (1972), *Peasant Marketing and Social Revolution in the State of La Paz, Bolivia*, Ph.D. thesis: McGill University.

Burke, M. (1970), 'Land Reform and its Effect upon Production and Productivity in the Lake Titicaca Region', *Economic Development and Cultural Change*, Vol.18, No.3, April.

Burke, M. (1971), 'Land Reform in the Lake Titicaca Region', in Malloy, J.M. and Thorn, R.S. (eds.), *Beyond the Revolution: Bolivia Since 1952*.

Calani G., E. (1996), *Pensamiento político ideológico campesino*, Litográfica "PAZ": La Paz.

Campfens, D. (1998), 'New Technologies Boost Farming', *Bolivian Times* (La Paz), 1 October.

Cárdenas Conde, V.H. (1994), 'Many Peoples, One Earth, One Nation', *Grassroots Development*, Vol. 17, No.2.

Carter, W.E. (1964), *Aymara Communities and the Bolivian Agrarian Reform*, University of Florida Press: Gainesville.

Carter, W.E. (1971a), *Bolivia: A Profile*, Praeger Publishers: New York.

Carter, W.E. (1971b), 'Revolution and the Agrarian Sector', in Malloy, J.M. and Thorn, R.S. (eds.), *Beyond the Revolution: Bolivia Since 1952*.

Chonchol, J. (1970), 'Eight Fundamental Conditions of Agrarian Reform in Latin America', in Stavenhagen, R. (ed.), *Agrarian Problems and Peasant Movements in Latin America*.

Christian Aid (1998), *Forever in your debt? Eight poor nations and the G-8: Millenium debt relief for eliminating poverty*, Christian Aid: London.

Clark, R.J. (1968), 'Land Reform and Peasant Market Participation on the Northern Highlands of Bolivia', *Land Economics*, Vol.XLIV, May.

Clark, R.J. (1969), 'Problems and Conflicts over Land Ownership in Bolivia', *Inter-American Economic Affairs*, Vol.22, No.4. Spring.

Clark, R.J. (1970), *Land Reform in Bolivia*, USAID/Bolivia and Land Tenure Center, University of Wisconsin: Madison.

Clark, R.J. (1971), *Reforma Agraria e Integración Campesina en la Economia Boliviana*, SNRA: La Paz.

Coad, M. (1992), 'Know-how of the ancients helps to feed Bolivia's poor', *The Guardian* (UK), 20 October.

Conyers, D. (1982), *An Introduction to Social Planning in the Third World*, John Wiley and Sons: Chichester, UK.

Cox and Kings Travel Ltd. (1998), *Summer Travel Digest*: London.

Cramer, M. (1996), 'Indigenous Communities in Peril', *Bolivian Times* (La Paz), 26 September.

Dejoux, C. and Iltis, A. (eds.) (1991), *El Lago Titicaca: Síntesis del conocimiento limnológico actual*, ORSTOM and HISBOL: La Paz.

Denevan, W.M. (1985), 'Peru's Agricultural Legacy: Ancient methods may be useful in reviving today's food production', *Focus*, April.

Denevan, W.M., Smith, C.T. and Hamilton, P. (1968), 'Ancient ridged fields in the region of Lake Titicaca', *Geographical Journal*, Vol.134, No.3.

Dickenson, J., Gould, B., Clarke, C., Mather, S., Prothero, M., Siddle, D., Smith, C. and Thomas-Hope, E. (1996, 2nd ed.), *A Geography of the Third World*, Routledge: London.

Dion, H.G. (1950), *Agriculture in the Altiplano of Bolivia*, Development Paper No. 4, FAO: Washington.

Dorner, P. (1966), *Land Tenure Reform and Agricultural Development in Latin America*, Land Tenure Center, University of Wisconsin: Madison.

Dorner, P. (1992), *Latin American Land Reforms in Theory and Practice: A Retrospective Analysis*, University of Wisconsin Press: Madison.

Dunkerley, J. (1984), *Rebellion in the Veins: Political Struggle in Bolivia 1952-1982*, Verso Editions: London.

Eicher, C.E. and Staatz, J.M. (eds.) (1990), *Agricultural Development in the Third World*, John Hopkins University Press: Baltimore.

Ellis, F. (1993, 2nd ed.), *Peasant Economics: Farm households and agrarian development*, Cambridge University Press: Cambridge.

Erasmus, C.J. (1967), 'Upper Limits of Peasantry and Agrarian Reform: Bolivia, Venezuela and Mexico Compared', *Ethnology*, 6.

Feder, E. (1971), *The Rape of the Peasantry: Latin America's Landholding System*, Anchor Books: New York.

Flores, E. (1970) 'The Economics of Land Reform', in Stavenhagen, R. (ed.), *Agrarian Problems and Peasant Movements in Latin America*.

Frank, A.G. (1971) *Capitalism and Underdevelopment in Latin America*, Penguin: Middlesex, UK.

Galeano, E. (1992), 'Behind the Mask of History', *500 Years of Resistance*, Issue One, Latin America House: London.

García, A. (1970), 'Agrarian Reform and Social Development in Bolivia', in Stavenhagen, R. (ed.), *Agrarian Problems and Peasant Movements in Latin America*.

Gelber, G. (ed.) (1992), *Poverty and Power in Latin America after 500 Years*, CAFOD: London.

Gonzales, M. (1992), 'From Columbus to Bush - and back', *500 Years of Resistance*, Issue Four, Latin America House: London.

Goodrich, C. (1971), 'Bolivia in Time of Revolution', in Malloy, J.M. and Thorn, R.S. (eds.), *Beyond the Revolution: Bolivia Since 1952.*

Graeff, P. (1974), *The effects of continued landlord pressure on the Bolivian countryside during the post-reform era: lessons to be learned*, Land Tenure Center, University of Wisconsin: Madison.

Grant, J. (1998), 'Indigenous people fight for forest', *Latinamerica Press* (Lima), 12 February.

Grant, J. (1998), 'Migrants flood El Alto', *Latinamerica Press* (Lima), 16 July.

Grillo, E. and Rengifo, G. (1990), *Agricultura y Cultura en los Andes*, HISBOL/PRATEC: La Paz.

Guarachi, P. (1992), 'A new agrarian order for the new generation', *Bolivia Bulletin* (La Paz), December.

Gwynne, R.N. and Kay, C. (1997), 'Agrarian Change and the Democratic Transition in Chile: An Introduction', *Bulletin of Latin American Research*, Vol.16, No. 1, January.

Heath, D.B. (1959), 'Land Reform in Bolivia', *Inter-American Economic Affairs*, 12.

Heath, D.B., Erasmus, C.J. and Buechler, H.C. (1969), *Land Reform and Social Revolution in Bolivia*, Praeger: New York.

Hemming, J. (1970), *The Conquest of the Incas*, Macmillan: London.

Huizer, G. (1973), *Peasant Rebellions in Latin America*, Penguin Books Ltd.: UK.

Inter-American Development Bank (1997), *Annual Report*, Washington.

Jeffrey, P. (1997), 'Landowners gripe about bishops', *Latinamerica Press* (Lima), 24 April.

Jenkins, R. (1997), 'Trade Liberalisation in Latin America: the Bolivian Case', *Bulletin of Latin American Research*, Vol.16, No. 3, September.

Kay, C. (1997), 'Globalisation, Peasant Agriculture and Reconversion', *Bulletin of Latin American Research*, Vol.16, No.1, January.

Klein, H.S. (1992, 2nd ed.), *Bolivia: The Evolution of a Multi-Ethnic Society*, Oxford University Press: New York.

Koning, H. (1991), *Columbus: His Enterprise*, Latin America Bureau: London.

Ley INRA (1997), *Colección Jurídica Librería Universitaria*: Cochabamba.

LIDEMA (1992), *El Estado del Medio Ambiente en Bolivia*, LIDEMA: La Paz.

Lindqvist, S. (1972), *The Shadow: Latin America faces the Seventies*, Penguin Books: UK.

Lindqvist, S. (1974), *Land and Power in South America*, Penguin Books: UK.

Lockhart, J. (1969), 'Encomienda and Hacienda: The Evolution of The Great Estate in the Spanish Indies', *Hispanic American Historical Review*, August.

Loza, E. Articles in *Bolivian Times* (La Paz): 'LIDEMA keeps Bolivia green into the 21st Century' (5 September 1996); 'Desertification and Droughts

Increase' (30 January 1997); 'Why Bolivia Can't Feed Itself' (16 October 1997); 'Farmers suffer through unpredictable El Niño' (5 February 1998).

Lumbreras, L.G. (1991), 'Misguided Development', *NACLA Report on the Americas*, Vol. XXIV, No. 5, February.

MacDonald, N. (1992), *The Andes: A Quest for Justice*, OXFAM: UK.

Malloy, J.M. and Thorn, R.S. (eds.) (1971), *Beyond the Revolution: Bolivia Since 1952*, University of Pittsburgh Press.

Manz, T. and Jara, B. (1996), *La Agricultura Sostenible y el Medio Rural en Bolivia: Comentarios a una Propuesta*, Artes Gráficas: La Paz.

Mason, J.A. (1957), *Ancient Civilizations of Peru*, Penguin Books: Edinburgh.

Materne, Y. (ed.) (1980), *The Indian Awakening in Latin America*, Friendship Press: New York.

McBride, G.M. (1921), 'The Agrarian Indian Communities of Highland Bolivia', *American Geographical Society Research Series*, No. 5, Oxford University Press: New York.

McEwen, W.J. (1975), *Changing Rural Society: A Study of Communities in Bolivia*, Oxford University Press: New York.

McFarren, P. (1996), 'The sacred spots of Lake Titicaca and Copacabana turn to rubbish', *Bolivian Times* (La Paz), 28 November.

Ministerio de Hacienda (1996), *Estrategia para la transformación productiva del agro*, La Paz.

Miranda S., R.P.P. (1970), *Diccionario Breve: Castellano-Aymara, Aymara-Castellano*, La Paz.

Montes de Oca, I. (1992), *Sistemas de Riego y Agicultura en Bolivia*, MACA/CIIR: La Paz.

Montoya, R. (1990), 'Latin America: 500 years of conquest', *Latinamerica Press* (Lima), 26 April.

Morris, A. (1987, 3rd ed.), *South America*, Hodder and Stoughton: London.

Murra, J.V. (1975), *Formaciones económicas y políticas del mundo andino*, Instituto de Estudios Peruanos: Lima.

Odell, P.R. and Preston, D.A. (1973), *Economies and Societies in Latin America: A Geographical Interpretation*, John Wiley and Sons Ltd: Budapest.

Osborne, H. (1964, 3rd ed.), *Bolivia: A Land Divided*, Oxford University Press: London.

OXFAM (1972), *Proyecto Oxfam Extensionistas Cooperativas: Informe de Actividades*: Oxford.

Parodi I.,A. (1995), *El Lago Titicaca: Sus Características Fisicas y Sus Riquezas Naturales, Arqueológicas*, Arequipa: Peru.

Patch, R. (1961), 'Bolivia: The Restrained Revolution', *The Annual of the American Academy of Political and Social Sciences*, Vol.334.

Paz Estenssoro, V. (1956), *Mensaje del Presidente de la República al H. Congreso Nacional*, MNR: La Paz.

Prebisch, R. (1962), 'Aspects of the Alliance', in Dreier, J.C. (ed.), *The Alliance for Progress - Problems and Perspectives*, Baltimore.

Preston, D.A. (1969), *A Survey of Land Tenure and Land Use in Peasant Communities on the Central Altiplano of Bolivia*, CIDA Research Paper No. 6, Land Tenure Center, University of Wisconsin: Madison.

Quarrie, J. (ed.) (1992), *Earth Summit '92: The United Nations Conference on Environment and Development: Rio de Janeiro 1992*, The Regency Press Corporation: London.

Quispe, C. (1988), 'Native theologian calls for new pastoral approach that reflects Aymara world view', *Latinamerica Press* (Lima), 14 April.

Salles, V. (1994), 'Campesino widows demand land, justice', *Latinamerica Press* (Lima), 27 January.

Serrano Torrico, S. (ed.) (1993), *Ley de Reforma Agraria: Ley de Reforma Urbana*, Serrano: Cochabamba.

Smith, C.T. (1971), 'The Central Andes', in Blakemore, H. and Smith, C.T. (eds.), *Latin America: Geographical Perspectives*.

Sobrino, J. (1992), *500 Years: Structural Sin and Structural Grace: Reflections for Europe from Latin America*, Address given in Salford Cathedral, UK, 21 March.

Solón, P. (1995), *La Tierra Prometida: un aporte al debate sobre las modificaciones a la legislación agraria*, CEDOIN: La Paz.

Solón, P. (1997), *Horizontes sin tierra? Análisis crítico de la Ley INRA*, CEDOIN: La Paz.

Soux, J.A. (1987), *Manual Agrícola*, Wayar and Soux Ltda: La Paz.

Stavenhagen, R. (ed.) (1970), *Agrarian Problems and Peasant Movements in Latin America*, Anchor Books: New York.

Swaney, D. and Strauss, R. (1992), *Bolivia: A Travel Survival Kit*, Lonely Planet Publications: Australia.

Sykes, C. (1998), 'Raised Hopes for Highland Farmers', *Bolivian Times* (La Paz), 30 July.

Tamayo, F. (1910), *Creación de la pedagogía nacional*, La Paz.

Tapia Vargas, G. (1994), 'La Agricultura en Bolivia', *Enciclopedia Boliviana*, Los Amigos del Libro: Cochabamba.

Thiesenhusen, W.C. (1995), *Broken Promises: Agrarian Reform and the Latin American Campesino*, Westview Press: Boulder, Colorado.

Trend, J.B. (1946), *Bolivar and the Independence of Spanish America*, Hodder and Stoughton: London.

United Nations ECLA (1951), *Development of Agriculture in Bolivia*, UN: Mexico City.

Urquidi, M.A. (1969), *Bolivia y su Reforma Agraria*, Cochabamba.

Vargas, M., Articles in *Bolivian Times* (La Paz): 'The Blue Mirror of the Altiplano is Contaminated' (19 December 1996); 'Quinua: Food for the 21st Century' (22 May 1997).

Vidal, J. (1997), 'The long march home', *The Guardian* (UK), 26 April.

Von Hagen, V.W. (1957), *Realm of the Incas*, The New English Library Ltd: London.

Wearne, P. (1996), *Return of the Indian: Conquest and Revival in the Americas*, Cassell with Latin America Bureau: London.

Whitbeck, R.H. and Williams, F.E. (1940), *Economic Geography of South America*, McGraw-Hill Series in Geography.

Whittemore, C. (1981), *Land for People: Land tenure and the very poor*, OXFAM: Oxford.

Wolf, E.R. (1955), 'Types of Latin American Peasantry', *American Anthropologist*, Vol.57, No. 3.

World Bank (1974), 'Land Reform', *Rural Development Series:* Washington.

World Population Data Sheet (1998), Population Reference Bureau: Washington.

Zondag, C.H. (1966), *The Bolivian Economy 1952-65: The Revolution and its Aftermath*, Praeger: New York.

Zoomers, A. (1997), 'Titling land in Bolivia: Searching for a redefinition of tenure regimes', in Van Naerssen, T., Rutten, M. and Zoomers, A. (eds.) *Diversity of Development: Essays in Honour of Jan Kleinpenning*, Van Gorcum: Assen.

Printed and bound by CPI Group (UK) Ltd, Croydon, CR0 4YY

22/10/2024

01777628-0006